Moby-Dick

A Picture Voyage

Library of Congress Cataloging–in–Publication Data

Melville, Herman, 1819-1891
   Moby-Dick : a picture voyage : an abridged and illustrated edition of the original classic / by Herman Melville ; edited by Tamia A. Burt, Joseph D. Thomas, Marsha L. McCabe ; with illustrations from the New Bedford Whaling Museum.
   p. cm.
  ISBN 0-932027-68-7 (pbk.) -- ISBN 0-932027-73-3 (Cloth)
  1. Melville, Herman, 1819-1891. Moby Dick--Illustrations. 2. Sea stories, American--Illustrations. 3. Whaling ships--Pictoirial works. 4. Whaling--Pictorial works. 5. Whales--Pictorial works. I. Burt, Tamia A. II. Thomas, Joseph D. III. McCabe, Marsha. IV. Title.
PS2384.M6 A36 2002
813'.3--dc21
                              2002009311

© 2002 by Spinner Publications, Inc. All rights reserved.
Printed in the United States of America
Spinner Publications, Inc., New Bedford, MA 02740

Moby-Dick

A Picture Voyage

An Abridged and Illustrated Edition of the Original Classic

by

Herman Melville

Edited by

Tamia A. Burt, Joseph D. Thomas, Marsha L. McCabe

with illustrations from The New Bedford Whaling Museum

Acknowledgments / Credits

Naturally, no serious book concerning the American whaling industry can be done without interaction with the New Bedford Whaling Museum. We are grateful to Director Anne Brengle and Director of Programs Lee Heald for their support. We are especially grateful to the Museum's library staff, particularly Assistant Librarian Laura Pereira and Librarian Michael Dyer, for their energy and helpfulness, and to Collections Manager Mary Jean Blasdale, Curator Michael Jehle, volunteer Irwin Marks, Emeritus Director Richard Kugler, and Photo Archivist Michael Lapides.

When we began work on this project, The Kendall Whaling Museum was an independent entity in Sharon, Massachusetts, and we were fortunate enough to receive the gracious assistance and eminent knowledge of the Kendall's Director, Stuart M. Frank, who provided detailed information on all images from the Kendall Collection. We are also grateful to the Kendall's entire staff in Sharon. The Kendall Museum has since merged with the New Bedford Whaling Museum and we have credited all images from the Kendall as "Kendall Collection, NBWM."

There are two families that provided use of their fabulous photographic collections: The Kingman/Greenhalgh family of New Bedford, Dartmouth and Nantucket, and the William R. Hegarty family. Special thanks for courtesy usage from the MIT Museum, the Millicent Library in Fairhaven, R. R. Donnelley & Sons Company, Arthur Moniz and Ted Holcombe.

Thanks to the Mayor of New Bedford, Frederick M. Kalisz, Jr., for his support.

Editing

Tamia Burt
Joseph D. Thomas
Marsha L. McCabe

Design / Art Editor

Joseph D. Thomas

Electronic Imaging

Jay Avila

Production

Andrea V. Tavares, copy-editing
Anne J. Thomas, copy-editing
Milton P. George, marketing
Jim Grasela, copy-editing
Ruth Caswell, research
Claire Nemes, marketing
Kendall Mattos, research
John K. Robson, photography
Seth Beall, photography

Collections

Arthur Moniz Studios
Edwards Family Collection
Forbes Collection of the MIT Museum
William R. Hegarty Collection
T. E. Holcombe
Kendall Collection, NBWM
Kingman Family Collection
Millicent Library, Fairhaven
National Archives
New Bedford Free Public Library
New Bedford Whaling Museum
Peabody Salem Museum
R. R. Donnelley & Sons Company
Rosenbach Museum, Philadelphia

Support provided by

The Massachusetts Cultural Council
The City of New Bedford

Contents

Preface		6
Introduction		7

Chapter

1	Loomings	13
2	The Carpet-Bag	16
3	The Spouter-Inn	20
4	The Counterpane	27
5, 6	Breakfast · The Street	28
7, 8	The Chapel · The Pulpit	32
9	The Sermon	34
10, 11	A Bosom Friend · Nightgown	36
12	Biographical	37
13	Wheelbarrow	38
14	Nantucket	48
15	Chowder	49
16	The Ship	51
18	His Mark	60
20	All Astir	61
21	Going Aboard	63
22	Merry Christmas	64
23, 24	The Lee Shore · The Advocate	66
26, 27	Knights and Squires · Knights and Squires	68
28	Ahab	71
29	Enter Ahab; to him, Stubb	72
30, 31	The Pipe · Queen Mab	74
32	Cetology	75
33, 34	The Specksynder · The Cabin Table	78
35	The Lookout	80
36	The Quarter-Deck	81
37	Sunset	84
41	Moby Dick	85
42	The Whiteness of the Whale	88
43, 44	Hark! · The Chart	90
45	The Affadavit	91
47	The Mat Maker	92
48	The First Lowering	93
49	The Hyena	96
50	Ahab's boat and crew. Fedallah	98
51	The Spirit Spout	99
52	The Albatross	101
53	The Gam	102
55	Of the Monstrous Pictures of Whales	104
56	Of the Less Erroneous Pictures of Whales	105
57	Of Whales in Paint; In Teeth; In Wood	110
58, 60	Brit · The Line	111
61	Stubb Kills a Whale	114
62	The Dart	116
63	The Crotch	117
64	Stubb's Supper	118
65	The Shark Massacre	121
67, 68, 69	Cutting In · The Blanket · The Funeral	122
70	The Sphynx	124
72	The Monkey Rope	125
73	Stubb and Flask Kill a Right Whale	126
74	The Sperm Whale's Head—Contrasted View	128
75	Right Whale's Head—Contrasted View	129
76	The Battering-Ram	131
77	The Great Heidelburgh Tun	132
78	Cistern and Buckets	133
81	The Pequod Meets the Virgin	135
84	Pitchpoling	141
85	The Fountain	142
86	The Tail	143
87	The Grand Armada	144
88	Schools and Schoolmasters	149
89	Fast-fish and Loose-fish	150
91, 92	The Pequod Meets the Rose-bud · Ambergris	151
93	The Castaway	154
94	A Squeeze of the Hand	155
95	The Cassock	157
96	The Try-works	158
98	Stowing Down and Clearing Up	161
100	Leg and Arm. The Pequod, of Nantucket	164
101	The Decanter	167
102	A Bower in the Arsacides	168
103	Measurement of the Whale's Skeleton	169
105	Does the Whale's Magnitude Diminish?—Will he Perish?	170
106, 107, 108	Ahab's Leg · The Carpenter · Ahab and the Carpenter	171
109	Ahab and Starbuck in the Cabin	172
110	Queequeg in his Coffin	173
111	The Pacific	174
112, 113	The Blacksmith · The Forge	180
114	The Gilder	182
115	The Pequod meets the Bachelor	183
116	The Dying Whale	184
117	The Whale Watch	185
118	The Quadrant	186
119	The Candles	187
123	The Musket	189
124	The Needle	191
126	The Life Buoy	192
128	The Pequod Meets the Rachel	194
131	The Pequod Meets the Delight	196
132	The Symphony	197
133	The Chase—First Day	199
134	The Chase—Second Day	204
135	The Chase—Third Day	208
	Epilogue	215

Index		217
Bibliography		221

Preface

If you're looking for a good adventure story but untill now have been too intimidated by *Moby-Dick's* reputation to read it, this abridgement is for you. We've made this great novel more accessible so you can read and enjoy it without having to keep handy an encyclopedia or a Melville scholar. Yet *Moby-Dick: A Picture Voyage* maintains the plot and has kept much of the symbolism, allegory, humor and philosophy.

But alas, there are many abridged editions of *Moby-Dick* already in print. Thus we join an ever-expanding cadre of publishing sharks feeding off the carcass of Melville's leviathan. Poor Herman. He never lived to see the glory of his work, but he always believed in his mission.

The body text of this work is from the 1930 edition of *Moby-Dick* published by R.R. Donnelley & Sons Company. Though we have edited chapters and shortened a few paragraphs, we have not tampered with the author's sentence structure or grammatical style. The marathon sentences, peculiar punctuation and archaic spelling are pure Melville.

We've made this version unique by heavily illustrating it with a treasury of images that give an accurate account of Ishmael's great adventure and Melville's real-life whaling experience. We have transformed the book into a storyboard animated with authentic photographs that thoroughly portray the whaling industry and culture, as well as places visited by Ishmael and Melville.

We've included original whaler artwork such as logbook and journal paintings, scrimshaw, etchings and illustrations. You'll find drawings and engravings from old books and journals depicting nineteenth-century whaling. We have used canvas paintings and watercolors by master painters; and panoramas, commercial art, signage, ephemera, folk art, artifacts, maps and other images that faithfully represent the whaling culture during Melville's time and later.

Illustrating *Moby-Dick* seems so natural. The story fills our imaginations with a torrent of visual realities, and the descriptions of the day-to-day rigors of whaling have the ring of authenticity because of Melville's own experiences aboard the whaler *Acushnet*. Still, the task was daunting. After all (in the words of Ishmael), "To produce a mighty book, you must choose a mighty theme." A mighty theme indeed—Herman Melville and *Moby-Dick*.

In *Moby-Dick: A Picture Voyage*, we have tried to keep the story short and simple, yet true to the essence of the novel. First-time readers will discover Melville's sense of humor, his extensive knowledge of whaling and whale men, and his advocacy of the whale 150 years before "animal rights" and conservationism were recognized ideas: "the moot point is, whether Leviathan can long endure so wide a chase, and so remorseless a havoc; whether he must not at last be exterminated from the waters, and the last whale, like the last man, smoke his last pipe, and then evaporate himself in his last puff."

Melville was also a prophet, making astonishingly accurate predictions about the future of man. Though deploring the "butchering sort of business of whaling," Melville acknowledged how whalers and whale men changed the world: "If that double-bolted land, Japan, is ever to become hospitable, it is the whale-ship alone to whom the credit will be due; for already she is on the threshold." This prediction would come true within months of the release of *Moby-Dick* in 1851 (learn the story of John Manjiro, castaway rescued by an American whaleship).

But mostly you'll discover Melville's genius as a writer. He describes so vividly the sensations, emotions, and events of a whaling trip that you'll almost see, smell, and touch them for yourself.

Moby-Dick: A Picture Voyage is not a substitute for the original. You should still read an unabridged version of *Moby-Dick* so as not to miss the power and depth of the original text—no abridgement can ever be an improvement over this American classic.

—The Editors

Introduction

by Laurie Robertson-Lorant

In 1840, twenty-one-year-old Herman Melville arrived in New Bedford determined to sign on a whaleship bound for the Pacific. It was not a decision his family wholeheartedly approved. A grandson of two Revolutionary War heroes was expected to serve aboard a naval frigate, not sail before the mast on a "blubber-hunter." Most genteel people assumed whalemen were fugitives and felons, and quite often, they were. Several cousins who served as midshipmen in the Navy or went whale-hunting came to a bad end, usually as a result of overindulgence in alcohol. Melville, however, lived to tell the tale—several tales, in fact.

Moby-Dick, his masterpiece, is a great American epic on the order of *Gilgamesh*, *The Lusiads* or *The Odyssey*. The hunt for the white whale resonates with the power of ancient myth. Drawing on many sources, conscious and unconscious, *Moby-Dick* combines Native American legends, frontier tall tales and hunting stories with Biblical allusions, Rabelaisian scatology, Shakespearean poetry, sailor yarns and chanteys, Oriental philosophy, Puritan sermons, whaling lore, philosophical meditations and spiritual speculations. In addition to all this, *Moby-Dick* is a multicultural adventure.

Herman Melville, 1870. — *Painting by J. O. Eaton.*

Sounding a dramatic opening chord, Melville's narrator addresses the reader in the imperative voice. "Call me Ishmael," he commands, a Biblical reference that identifies him as an outcast. A schoolteacher with an empty purse and a "damp, drizzly November in [his] soul," he takes to sea, he tells us, to avoid committing suicide.

Arriving in New Bedford, he searches through "blocks of blackness" for a place to stay, at one point falling face-first into an ash-box on the porch of a church. He opens the door and peers in, only to see a sea of black faces turn towards him, and a preacher is talking about "the blackness of darkness." Ishmael imagines he has blundered upon the "black parliament in Tophet." Melville's joke is that Ishmael's soot-smudged white face must look very strange to the congregation of a Negro church.

He ends up at Peter Coffin's Spouter-Inn, where he is told he must share a bed with a "dark-complexioned" harpooneer. When a strange-looking fellow who looks

The town of New Bedford from the Fairhaven shore, 1839. *This is what Melville saw as the* Acushnet *pulled away from the wharf in Fairhaven in January 1842. — Woodcut drawing by John W. Barber. From* Historical Collections…of Every Town in Massachusetts. *1841.*

National Archives

as though he's been in a fight strolls into the room, his instinct is to recoil in fear, but he soon realizes his roommate is tattooed. Resolving to ignore such superficial markers of difference, he concludes, "It's only his outside, … a man can be honest in any sort of skin." Even when Queequeg scares him half to death by jumping into bed with a tomahawk between his teeth, Ishmael gives him the benefit of the doubt: "The man's a human being just as I am; he has just as much reason to fear me, as I have to be afraid of him. Better sleep with a sober cannibal than a drunken Christian."

Soon, the two men fall asleep as comfortably as "a married pair" in their "hearts' honeymoon," and Ishmael awakens to find Queequeg's arm thrown over him "in the most loving and affectionate manner." In the brotherly embrace of this cosmopolitan cannibal, the spiritually orphaned Ishmael feels his stiff Presbyterian prejudices "melting" and his misanthropy and despair dissolving. "No more my splintered heart and maddened hand were turned against the wolfish world. This soothing savage had redeemed it."

Queequeg is a hybrid, a multicultural icon, a collage. He carries an embalmed head on his belt in the fashion of the Maori warriors, wears a Native American wampum belt, smokes a tomahawk peace pipe, prays to an African idol named Yojo, and appears to observe a Ramadan. The next day, the "bosom friends" sign on the whaler *Pequod* despite the warnings of a crazy prophet named Elijah who says they are doomed if they agree to serve with Captain Ahab, whom he calls "Old Thunder."

Ishmael and Ahab embody antithetical attitudes toward life. Whereas Ahab lusts to revenge himself on the white whale, Ishmael is drawn to the open ocean by "the overwhelming idea of the great whale" who entices him through the "floodgates of the wonder-world" to "wild and distant seas" where he experiences "all the attending marvels of a thousand Patagonian sights and sounds." Looking to Nature for revelations of the Truth, Ishmael gains reverence for the whale and learns to savor moments of peace and transcendence rather than searching obsessively for fixed belief. Although for a time Ishmael becomes caught up in Ahab's monomania, in the end, he breaks free to achieve an open-mindedness that moves him from suicidal despair to spiritual rebirth.

Born in New York City, a wind's rush from the Battery on Pearl Street—so named in colonial days for its iridescent surface of crushed oyster shells—Melville grew up amid dingy warehouses and wharves and "old-fashioned coffeehouses" with "sunburnt sea-captains going in and out, smoking cigars, and talking about Havana, London, and Calcutta." He liked to spend Sunday afternoons in his father's library, browsing through the two large green portfolios of prints his father had brought back from France, especially the one that showed whaleboats pursuing a whale stuck full of harpoons, pouring blood into the sea. Every Sunday, his father drew his chair up beside the coal fire and told the children stories about the "monstrous waves at sea, mountain high; of the masts bending like twigs; and all about Havre, and Liverpool." Listening to these stories, young Herman "fell into long reveries about distant voyages and travels." He read *The Thousand and One Nights* and *The Travels of Marco Polo* and yearned to explore "remote and barbarous countries" and return with "dark and romantic sunburnt cheeks." Early proximity to New York Harbor and the influence of maritime literature and art, together with his later shipboard experiences, combined to produce *Typee, Omoo, Mardi, Redburn, White-Jacket* and *Moby-Dick*.

Melville's father, Allan Melvill (Herman added the final 'e' later) shared his wife's obsession with moving up socially. He lived on borrowed money, robbing Peter to pay Paul until he went bankrupt and had to flee New York City for Albany, his wife's hometown. Not long after the move, he contracted pneumonia from overexposure to frigid temperatures, and, within a fortnight, Allan Melvill was dead. The shock twelve-year-old Herman experienced was profound. He left school and worked for a bank to help his widowed mother make ends meet. This job lasted long enough to convince him he did not want to end up "pent up in lath and plaster" and "clinched to desks." He much preferred pitching hay on Uncle Thomas' Berkshire farm, but that was only a summer job.

When Melville was sixteen, his mother moved the family across the river to Lansingburgh, New York, a bustling river port whose shipyards produced almost 300 oceangoing vessels between 1780 and 1830. From their home on River Street, Melville could see oceangoing ships in various stages of construction, and during the winter, when the river was to too low to float the larger ships, he and his friends could play in the huge storage tunnels built under the street. Lansingburgh was an enchanted village for a boy with a vivid imagination.

Struggling to raise eight youngsters with very little money, Maria Gansevoort Melville relied on her brother Peter for advice. At his suggestion, Herman studied surveying at Lansingburgh Academy with a view toward getting a job on the Erie Canal, but by the time he finished his course, there were no jobs. Melville then tried teaching school near Pittsfield, not far from his uncle's farm, but he found the disciplining of unruly farm boys burdensome and demoralizing.

Living amidst ships and sailors had whetted Herman's appetite for the sea. In the summer of 1839, he signed on as a cabin-boy aboard a packet-ship bound for Liverpool. As the wind caught the sails, the ship "gave a sort of bound like a horse" and "went plunging along, shaking off the foam from her bows like the foam from a bridle-bit," and he found himself entering the "wide blank" of the Atlantic Ocean.

Whatever anxiety and loneliness Melville may have felt as all sight of land dropped away and the ocean surrounded the small ship, his worries were offset by the pleasure he felt as the ship surged under him. "Every mast and timber seemed to have a pulse in it that was beating with life and joy; and I felt a wild exulting in my own heart, and felt as if I would be glad to bound along so round the world."

Like all "lubbers," he soon found himself assigned such unpleasant tasks as cleaning out pig pens and chicken coops and swabbing the head, and such dangerous ones as scampering up the rigging to reef the sails in a storm. As a greenhorn, he had to follow orders barked at him by ruffians in a jargon that was foreign to his ear:

"What did I know, for instance, about *striking a top-gallant-mast*, and sending it down on deck in a gale of wind? Could I have *turned in a dead-eye*, or in the approved nautical style have *clapped a seizing on the main-stay*? What did I know of *passing a gammoning, reeving a Burton, strapping a shoe-black, clearing a foul hawse*, and innumerable other intricacies."

This first ocean voyage cemented the connection between splicing ropes and splicing words that enabled Melville to capture "the poetry of salt water."

At the end of the summer, he returned to his mother's house and wrote some love poems and gothic stories for the local newspaper, but there was no money in it, so he went back to teaching, this time at East Greenbush and Schodack Academy. Unfortunately, the school ran out of money and was forced to close. Disheartened and restless, he decided to go West to visit Uncle Thomas, who had moved his family to Galena, Illinois. Although the professed object of the trip was to find work on the frontier, Melville and his friend Eli M. Fly saw the prairies and St. Anthony's Falls and came home empty-handed.

His mother's financial situation had worsened, so Melville decided to go to sea again. Just before Christmas, 1840, he shouldered his duffel bag, exchanged farewells with his family and friends, and headed for America's premier whaling port, New Bedford. Accompanied by his brother Gansevoort, he signed an affidavit of American citizenship before crossing the river to Fairhaven.

There he signed on the *Acushnet*, the newest addition to America's 600-vessel whaling fleet, built at the Barstow Yard in Mattapoisett. The 359-ton-square-rigged ship had two decks and three masts, each with a crow's nest where the man on the lookout for whales would stand. The ship was so new it had not even been registered when Melville signed the papers.

The ship's manifest shows that "Herman Melville: birthplace, New York; age 21; height 5 feet 9 1/2 inches; complexion dark; hair, brown," was given an $84 advance against his pay to equip himself with the necessities for a four-year voyage. With the help of his brother, he purchased an oilskin suit, red flannel shirt, duck trousers, a straw tick, pillow, blankets for his bunk, a sheath knife and fork, a tin spoon and tin plate, needles, thread and mending cotton, soap, a razor, a ditty bag and a large sea chest in which to store these and other "necessaries." Once Herman was officially registered and properly supplied, Gansevoort went on to Boston while Herman explored the seaport on his own.

Nationwide, by 1848, more than $70 million was invested in the whaling industry and 70,000 persons derived their livelihood from it. The industry's birthplace was Nantucket, and the small island off the southeast coast of Massachusetts was world-renowned in the trade. But by the early 19th century, after several of Nantucket's leading merchants moved to New Bedford, the small village on the Acushnet River quickly flourished and, by the mid-1850s, New Bedford was reputed to be the richest city, per capita, in the world. In 1857 alone, some 329 vessels (about half the American fleet)

Merchants Wharf, New Bedford, 1860s.

New Bedford Whaling Museum

valued at over $12 million and employing 12,000 seamen, left New Bedford for the whaling grounds.

New Bedford was a gritty, bustling port and also a beautiful city, thanks to the whaling industry. Up the hill on County Street stood grand houses and gardens built by whaling entrepreneurs; below these were more modest but stately homes of captains and merchants; down the street whaling-related industries and small businesses flourished on the waterfront.

During the week Melville waited for his ship to sail, New Bedford was blanketed by snow and the river was frozen over. Much was going on around the city. At the Hillman Shipyard, a new ship was being built, the *Charles W. Morgan*. At the Port Society, president Samuel Rodman was once again campaigning against the rum-sellers. Richard Henry Dana was in town lecturing on Shakespeare and the drama.

As Melville ambled along the waterfront, he saw sailors from around the world: Swedes with blonde hair and blue eyes, Cape Verdeans with black curly hair and ebony skin, and South Sea islanders covered with tattoos that looked like star charts from undiscovered galaxies.

On Sunday in New Bedford, Melville attended services at the Seaman's Bethel, where the Rev. Enoch Mudge preached to sailors of all faiths or no faith at all. This whalemen's chapel, dedicated in 1832, looked down toward the harbor, as though the vigilance of God's minister could guarantee the safe return of the fleet. On the chapel wall hung, then as now, marble cenotaphs commemorating sailors lost at sea, some washed overboard and drowned, some crushed between boats and flailing whales, some wrecked on remote reefs, all resting at the bottom of the sea. As Ishmael tells us, the black-lettered white marble tablets served as a "doleful" reminder of "the fate of whalemen who had gone before me."

Whaling was so hazardous that the odds were 2 to 1 against a man's returning from a voyage, and only one of America's original wooden whaling vessels survives today: the *Charles W. Morgan*, now berthed at Mystic Seaport in Connecticut. "For every drop of oil, at least one drop of blood," whalemen would boast, and Navy men and sailors in the merchant service would respond, "Better dead than shipped aboard a blubber-hunter for a four-year cruise."

Despite these dire warnings, whaling was an adventure not to be missed. Along with the thrill of the chase and the capture of great whales, it promised at least an overnight stop in the Marquesas en route to the Japanese whaling ground, and perhaps, a few hours of dalliance with half-naked maidens whose sexual favors were a form of hospitality. But for this story, one must read *Typee: a Peep at Polynesian Life*.

Melville's great adventure began on January 3, 1841, in arctic cold, when the *Acushnet* left from Fairhaven, under the command of Valentine Pease II of Edgartown. He had 26 shipmates including four Portuguese, three black Americans, one Scotsman, one Englishman and various white Americans of different nationalities. In the 1830s, most Americans who joined whaling crews tended to be New England farm and village boys. By the mid-1840s, more "packet rats"—rovers and drifters, drunkards and fugitives signed on and whalers became asylums for outcasts. Melville was fortunate in his shipmates, who were superior to the ordinary run of whaling crews.

"A whaleship was my Harvard and my Yale," he wrote.

Ishmael's voyage (and Melville's), unlike Ahab's mad quest to revenge himself on the White Whale, is a voyage of discovery. They encounter new cultures and learn to appreciate multiple points of view: first Queequeg's, then Ahab's, then black Pip's, the views of the captains of the other ships the *Pequod* (and the *Acushnet*) meet on the high seas. At various times in the narrative, Ishmael even enters the consciousness of the great whales.

After 18 months on the *Acushnet* with an ill-tempered captain, Herman Melville deserted with his friend and shipmate, Toby Greene, at Nukaheva, one of the Marquesas islands. Walking over volcanic mountains and steep gorges, they came to the fertile valley of the Typees, rumored to be cannibals. Melville, who was suffering from fever, chills and an infected leg, persuaded the Polynesians to send Toby for medicine. Toby never returned because he was shanghaied by a whaler short of crew. Though the Typees were hospitable, Melville worried that they might be fattening him up, so he escaped to the southern coast of the island where he was picked up by an old whaling ship from Sydney, which took him to Tahiti. He and 11 crewmates refused further duty and he ended up in a makeshift prison, the "Calabooza", where he spent several months beachcombing. From Tahiti, Melville boarded the Nantucket whaleship *Charles and Henry*, then shipped for America as an enlisted man on the frigate *United States*.

After Melville arrived home from the Pacific in October, 1844, he began to write about his experiences in the South Seas. Two years later, he published *Typee: A Peep at Polynesian Life*, which was a huge success; he quickly followed it with a sequel, *Omoo*, a picaresque version of his sojourn in Tahiti and the Sandwich Islands. The success of these first two novels enabled Melville to ask Lemuel Shaw, an old family friend, for the hand of his daughter Elizabeth. They married in 1847 and moved to a townhouse they shared with his brother Allan and his wife, Sophia. There he wrote three novels in two years.

By the time Melville began writing *Moby-Dick* in 1850, both Lizzie and Sophia Melville had given birth to sons, and *"The Whale,"* as he called his book, had to compete with babies teething in the heat. Eager to escape New York, he took his family to the Berkshires, where he met Nathaniel Hawthorne, whose friendship was to have a profound effect on *Moby-Dick*. In October, Melville purchased a farmhouse in Pittsfield. He and Lizzie, their four children and sometimes Herman's mother and two of his sisters, lived there for the next thirteen years.

Moby-Dick; or the Whale was published in 1851, drawing either disapproval or indifference from reviewers. In July, 1852, Melville's father-in-law, Lemuel Shaw, Chief Justice of the Massachusetts Supreme Court, invited Melville to accompany him on the circuit so he could meet "some of the gents at New Bedford & Nantucket." In New Bedford, they dined with future governor John Clifford, the Attorney General who had successfully prosecuted Harvard professor John Webster for the grisly murder of his colleague, George Parkman. They also visited the Arnold gardens (ancestor to Harvard's Arnold Arboretum) before sailing for Nantucket. There they dined with Maria Mitchell, the young woman to whom the King of Denmark had awarded a medal for discovering a comet with her father's telescope.

Between 1852 and 1856, Melville wrote *Pierre, or The Ambiguities*, a novel the critics hated, and stories and sketches for *Harpers' Monthly* and *Putnam's Magazine*. "Bartleby the Scrivener: A Story of Wall Street," "Benito Cereno" and "The Encantadas" are three of the finest of these prose works. Sadly, *Putnam's* ceased publication shortly after Melville finished his novel *The Confidence Man: His Masquerade* in 1856.

Exhausted and suffering from the effects of rheumatism and alcohol, Melville sailed for Europe and the Holy Land, having "pretty much made up his mind to be annihilated," according to Nathaniel Hawthorne, who was now U. S. Consul in Liverpool. Although Melville survived the trip, he once again returned to face a dire financial situation. He applied for a government job, but was unsuccessful, so he took to the lecture circuit. In February, 1858, his lectures brought him back to New Bedford. He spoke on the "Statues of Rome" at the Lyceum before an audience of 400, a small audience for a speech in those days. Unfortunately, he was not a polished enough orator to draw big crowds.

Returning home to Arrowhead, he concentrated on poetry, eventually publishing a volume of poems about the Civil War. In 1862, Melville's sister Catherine moved to New Bedford with her husband, John Chipman Hoadley, who was brought in to run the New Bedford Copper Company during the war. A poet as well as a

Hoadley residence, 1960s. *The home of Melville's sister, Catherine, at Madison and Orchard Street in New Bedford.* — Spinner Collection.

brilliant engineer, Hoadley was a sensitive and appreciative reader of Melville's poetry, and he and Melville became close friends. Hoadley lived at 100 Madison Street between 1862 and 1866, and it is likely Melville visited him several times.

In 1863, shortly after the Draft Riots, Herman and Lizzie and their four children moved back to New York City, and Melville found a job as an inspector for the Customs House. He earned $4 a day checking ships docked along the Hudson River for contraband, a post he held for 20 years. The job gave him financial security, a regular routine and time to slip away to write poetry on small slips of paper he kept in his jacket pocket. Parts of *Clarel*, his 18,000-line poem about a pilgrimage in the Holy Land, were written during breaks.

Although Melville stopped publishing after *Clarel* appeared in 1876, and had been almost forgotten when he died in 1891, he never stopped writing. Despite the deaths of older relatives and friends and the tragic suicide of his eldest son, Melville read and wrote voraciously, including three books of poetry which were privately printed, and *Billy Budd: An Inside Narrative*, which was not published until the Melville Revival of the 1920's.

Each book Melville wrote was in some sense an experiment with a new voice or form. He reinvented himself every time and, by the end of his life, he was no longer writing to stay alive; rather, he stayed alive so he could write. He died in his sleep, never dreaming that *Moby-Dick* would become one of the universally recognized classics of world literature.

Laurie Robertson-Lorant *is the author of "Melville: A Biography," (The University of Massachusetts Press, 1998), the only up-to-date, full-length, fully annotated, complete, one-volume biography of Melville. A graduate of Radcliffe College/Harvard University with a M. A. and Ph. D. from New York University. Dr. Robertson-Lorant is a teacher, scholar and published poet.*

MOBY-DICK;

OR,

THE WHALE.

BY

HERMAN MELVILLE,

AUTHOR OF
"TYPEE," "OMOO," "REDBURN," "MARDI," "WHITE-JACKET."

NEW YORK:
HARPER & BROTHERS, PUBLISHERS.
LONDON: RICHARD BENTLEY.
1851.

Chapter 1

Loomings

Call me Ishmael. Some years ago—never mind how long precisely—having little or no money in my purse, and nothing particular to interest me on shore, I thought I would sail about a little and see the watery part of the world. It is a way I have of driving off the spleen, and regulating the circulation. Whenever I find myself growing grim about the mouth; whenever it is a damp, drizzly November in my soul; whenever I find myself involuntarily pausing before coffin warehouses, and bringing up the rear of every funeral I meet; and especially whenever my hypos get such an upper hand of me, that it requires a strong moral principle to prevent me from deliberately stepping into the street, and methodically knocking people's hats off—then, I account it high time to get to sea as soon as I can. This is my substitute for pistol and ball. With a philosophical flourish Cato throws himself upon his sword; I quietly take to the ship. There is nothing surprising in this. If they but knew it, almost all men in their degree, some time or other, cherish very nearly the same feelings towards the ocean with me.

Say, you are in the country; in some high land of lakes. Take almost any path you please, and ten to one it carries you down in a dale, and leaves you there by a pool in the stream. There is magic in it. Let the most absent-minded of men be plunged in his deepest reveries—stand that man on his legs, set his feet a-going, and he will infallibly lead you to water, if water there be in all that region. Should you ever be athirst in the great American desert, try this experiment, if your caravan happen to be supplied with a metaphysical professor. Yes, as every one knows, meditation and water are wedded for ever.

Now, when I say that I am in the habit of going to sea whenever I begin to grow hazy about the eyes, and begin to be over conscious of my lungs, I do not mean to have it inferred that I ever go to sea as a passenger. For to go as a passenger you must needs have a purse, and a purse is but a rag unless you have something in it. Besides, passengers get seasick—grow quarrelsome—don't sleep of nights—do not enjoy themselves much, as a general thing;—no, I never go as a passenger; nor, though I am something of

View of Fairhaven from the New Bedford Shore, 1839. *The scene shows New Bedford Harbor around the time of Melville's arrival in 1841. "Let the most absent-minded of men be plunged in his deepest reveries—stand that man on his legs, set his feet a-going, and he will infallibly lead you to water, if water there be in all that region… Yes, as every one knows, meditation and water are wedded for ever."* – Chapter I. *Woodcut drawing by John W. Barber. From* Historical Collections…of Every Town in Massachusetts, *1841.*

National Archives

a salt, do I ever go to sea as a Commodore, or a Captain, or a Cook. I abandon the glory and distinction of such offices to those who like them. For my part, I abominate all honorable respectable toils, trials, and tribulations of every kind whatsoever. It is quite as much as I can do to take care of myself, without taking care of ships, barques, brigs, schooners, and what not.

No, when I go to sea, I go as a simple sailor, right before the mast, plumb down into the forecastle, aloft there to the royal mast-head. True, they rather order me about some, and make me jump from spar to spar, like a grasshopper in a May meadow. And at first, this sort of thing is unpleasant enough. It touches one's sense of honor, particularly if you come of an old established family in the land, the Van Rensselaers, or Randolphs, or Hardicanutes. And more than all, if just previous to putting your hand into the tar-pot, you have been lording it as a country schoolmaster, making the tallest boys stand in awe of you. The transition is a keen one, I assure you, from the schoolmaster to a sailor. But even this wears off in time.

What of it, if some old hunks of a sea-captain orders me to get a broom and sweep down the decks? Who aint a slave? Tell me that. Well, then, however the old sea-captains may order me about—however they may thump and punch me about, I have the satisfaction of knowing that it is all right; that everybody else is one way or other served in much the same way—either in a physical or metaphysical point of view, that is; and so the universal thump is passed round, and all hands should rub each other's shoulder-blades, and be content.

Again, I always go to sea as a sailor, because they make a point of paying me for my trouble, whereas they never pay passengers a single penny that I ever heard of. On the contrary, passengers themselves must pay. And there is all the difference in the world between paying and being paid. The act of paying is perhaps the most uncomfortable infliction that the two orchard thieves entailed upon us. But being paid,—what will compare with it? The urbane activity with which a man receives money is really marvellous, considering that we so earnestly believe money to be the root of all earthly ills, and that on no account can a monied man enter heaven. Ah! How cheerfully we consign ourselves to perdition!

Finally, I always go to sea as a sailor, because of the wholesome exercise and pure air of the forecastle deck. But wherefore it was that after having repeatedly smelt the sea as a merchant sailor, I should now take it into my

"View of New Bedford from the Fort near Fairhaven," 1845. *At this time, the whaling fleet consisted of 239 vessels and the city was said to be, per capita, the wealthiest in the United States. The whale fishery of the city reached its high point in capital, vessels and tonnage in 1857.*

Lithograph by Fitz Hugh Lane. *Spinner Collection.*

head to go on a whaling voyage; this the invisible police officer of the Fates, who has the constant surveillance of me, and secretly dogs me, and influences me in some unaccountable way—he can better answer than any one else. And, doubtless, my going on this whaling voyage, formed part of the grand programme of Providence that was drawn up a long time ago.

Though I cannot tell why it was exactly that those stage managers, the Fates, put me down for this shabby part of a whaling voyage, when others were set down for magnificent parts in high tragedies, and short and easy parts in genteel comedies, and jolly parts in farces—though I cannot tell why this was exactly; yet, now that I recall all the circumstances, I think I can see a little into the springs and motives which being cunningly presented to me under various disguises, induced me to set about performing the part I did, besides cajoling me into the delusion that it was a choice resulting from my own unbiased freewill and discriminating judgment.

Chief among these motives was the overwhelming idea of the great whale himself. Such a portentous and mysterious monster roused all my curiosity. Then the wild and distant seas where he rolled his island bulk; the undeliverable, nameless perils of the whale; these, with all the attending marvels of a thousand Patagonian sights and sounds, helped to sway me to my wish. With other men, perhaps, such things would not have been inducements; but as for me, I am tormented with an everlasting itch for things remote. I love to sail forbidden seas, and land on barbarous coasts. Not ignoring what is good, I am quick to perceive a horror, and could still be social with it—would they let me—since it is but well to be on friendly terms with all the inmates of the place one lodges in.

By reason of these things, then, the whaling voyage was welcome; the great flood-gates of the wonder-world swung open, and in the wild conceits that swayed me to my purpose, two and two there floated into my inmost soul, endless processions of the whale, and, mid most of them all, one grand hooded phantom, like a snow hill in the air.

Scrimshaw sailors, circa 1820s. Patriotism and free speech were popular themes among America's early seamen. The eagle perched on a bottle of "grog" holds a banner reading, "Free trade" and "Sailor's Rights"—themes from the War of 1812, during which sailors fought for the right to drink alcoholic beverages at sea. At the bottom of the sperm whale tooth, "Jack Contending for the Motto," indicates a competition is about to take place. — *Anonymous scrimshander. New Bedford Whaling Museum.*

Whaler's "Journel," 1856-60 (top right). Title page of the journal kept by Rodolphus W. Dexter of Tisbury, Martha's Vineyard aboard the bark Chili of New Bedford, Benjamin S. Clark, Master. Whalemen displayed unique artistry and craftsmanship in their journals and scrimshaw, providing a rare and invaluable record of their lives. — *Kendall Collection, New Bedford Whaling Museum.*

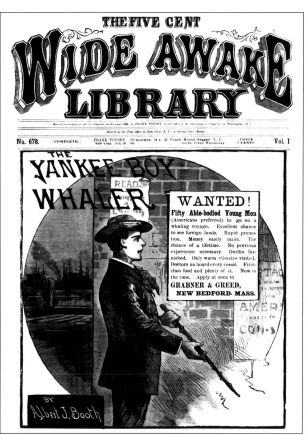

"Wide Awake Library," 1882. The cover of a pulp fiction magazine extols the romance of whaling with unrealistic promises of money, adventure and upward mobility. The need for able-bodied young men led to outlandish advertisements painting a rosy picture of one of the most exploitative and dirty jobs in America. Although "Yankee boys" were preferred, most crew members were foreign-born. Native sons knew better than to go whaling—unless driven to it as Ishmael was, when "the Fates put me down for this shabby part."

New Bedford Free Public Library

Chapter II

The Carpet-Bag

I stuffed a shirt or two into my old carpet-bag, tucked it under my arm, and started for Cape Horn and the Pacific. Quitting the good city of old Manhatto, I duly arrived in New Bedford. It was on a Saturday night in December. Much was I disappointed upon learning that the little packet for Nantucket had already sailed, and that no way of reaching that place would offer, till the following Monday.

As most young candidates for the pains and penalties of whaling stop at this same New Bedford, thence to embark on their voyage, it may as well be related that I, for one, had no idea of so doing. For my mind was made up to sail in no other than a Nantucket craft, because there was a fine, boisterous something about everything connected with that famous old island, which amazingly pleased me. Besides though New Bedford has of late been gradually monopolizing the business of whaling, and though in this matter poor old Nantucket is now much behind her, yet Nantucket was her great original—the place where the first dead American whale was stranded. Where else but from Nantucket did those aboriginal whalemen, the Red-Men, first sally out in canoes to give chase to the Leviathan? And where but from Nantucket, too, did that first adventurous little sloop put forth, partly laden with imported cobble-stones—so goes the story—to throw at the whales, in order to discover when they were nigh enough to risk a harpoon from the bowsprit?

"Just landed," New Bedford, 1860. — Harper's New Monthly Magazine.

"View from Acushnet Heights," painted circa 1868, depicting 1820s. *William A. Wall's idyllic interpretation of New Bedford's outskirts in the early 1800s differs markedly from Ishmael's: "Parts of her back country are enough to frighten one, they look so bony." Melville arrived in New Bedford during the frigid and snowy December of 1840. He sailed from (and probably boarded in) Fairhaven, which is depicted in the distance. This view gives a telescopic perspective, as it is taken from the top of the heights, about a mile from the waterfront.*

New Bedford Whaling Museum

Now having a night, a day, and still another night following before me in New Bedford, ere I could embark for my destined port, it became a matter of concernment where I was to eat and sleep meanwhile. It was a very dubious-looking, nay, a very dark and dismal night, bitingly cold and cheerless. I knew no one in the place. With anxious grapnels I had sounded my pocket, and only brought up a few pieces of silver,—So, wherever you go, Ishmael, said I to myself, as I stood in the middle of a dreary street shouldering my bag, and comparing the gloom towards the north with the darkness towards the south—wherever in your wisdom you may conclude to lodge for the night, my dear Ishmael, be sure to inquire the price, and don't be too particular.

With halting steps I paced the streets, and passed the sign of "The Crossed Harpoons,"—but it looked too expensive and jolly there. Further on, from the bright red windows of the "Sword-Fish Inn", there came such fervent rays, that it seemed to have melted the packed snow and ice from before the house, for everywhere else the congealed frost lay ten inches thick in a hard, asphaltic pavement,—rather weary for me, when I struck my foot against the flinty projections, because from hard, remorseless service the soles of my boots were in a most miserable plight. Too expensive and jolly, again thought I, pausing one moment to watch the broad glare in the street, and hear the sounds of the tinkling glasses within. But go on, Ishmael, said I at last; don't you hear? Get away from before the door; your patched boots are stopping the way. So on I went. I now by instinct followed the streets that took me waterward, for there, doubtless, were the cheapest, if not the cheeriest inns.

Such dreary streets! blocks of blackness, not houses, on either hand, and here and there a candle, like a candle moving about in a tomb. At this hour of the night, of the last day of the week, that quarter of the town proved all but deserted. But presently I came to a smoky light proceeding from a low, wide building, the door of which stood invitingly open. It had a careless look, as if it were meant for the uses of the public; so, entering, the first thing I did was to stumble over an ash-box in the porch.

Rising Sun Inn, Fairhaven, 1806–1846. Not quite the Spouter Inn, the Rising Sun Inn was established by Nicholas Taber in 1806 and located on Main Street just two blocks from where Melville's ship sailed. Because Fairhaven is where Herman's uncle Thomas shipped from on at least three of his four whaling voyages between 1835 and 1841, it's likely that Melville also boarded in Fairhaven. Perhaps he had occasion to visit the Rising Sun, where someone, "for their money, dearly sells the sailors deliriums and death." — Chapter 2. Millicent Library, Fairhaven.

View from lower Acushnet Heights, circa 1868. *This is the same view as the scene on opposite page. It depicts the northern fringe of a small, vibrant town that closely resembles what Melville saw in 1841. The Pearl Street railroad depot, built in 1840, is the long building at center facing broadside. Perhaps Melville arrived here, carpet-bag in hand, as he began his portentous voyage in search of whales.*

Kingman Family Collection

"Black Parliament." The Bethel A.M.E. Church on Kempton Street, established around 1842, was New Bedford's earliest known congregation for African Americans. New Bedford's eclectic, multi-cultural makeup was itself a curiosity in Melville's day. — Spinner Collection.

Ha! thought I, ha, as the flying particles almost choked me, are these ashes from that destroyed city, Gomorrah? But "The Crossed Harpoons," and "The Sword-Fish?"—this, then, must needs be the sign of "The Trap." However, I picked myself up and hearing a loud voice within, pushed on and opened a second, interior door.

It seemed the great Black Parliament sitting in Tophet. A hundred black faces turned round in their rows to peer; and beyond, a black Angel of Doom was beating a book in a pulpit. It was a negro church; and the preacher's text was about the blackness of darkness, and the weeping and wailing and teeth-gnashing there. Ha, Ishmael, muttered I, backing out, Wretched entertainment at the sign of "The Trap!"

Moving on, I at last came to a dim sort of light not far from the docks, and heard a forlorn creaking in the air; and looking up, saw a swinging sign over the door with a white painting upon it, faintly representing a tall straight jet of misty spray, and these words underneath—"The Spouter-Inn:—Peter Coffin."

Coffin?—Spouter?—Rather ominous in that particular connexion, thought I. But it is a common name in Nantucket, they say, and I suppose this Peter here is an emigrant from there. As the light looked so dim, and the place, for the time, looked quiet enough, and the dilapidated little wooden house itself looked as if it might have been carted here from the ruins of some burnt district, and as the swinging sign had a poverty-stricken sort of creak to it, I thought that here was the very spot for cheap lodgings, and the best of pea coffee.

It was a queer sort of place—a gable-ended old house, one side palsied as it were, and leaning over sadly. It stood on a sharp bleak corner, where that tempestuous wind Euroclydon kept up a worse howling than ever it did about poor Paul's tossed craft. Poor Lazarus there, chattering his teeth against the curbstone for his pillow, and shaking off his tatters with his shiverings, he might plug up both ears with rags, and put a corn-cob into his mouth, and yet that would not keep out the tempestuous Euroclydon.

But what thinks Lazarus? Can he warm his blue hands by holding them up to the grand northern lights? Would not Lazarus rather be in Sumatra than here? Would he not far rather lay him down lengthwise along the line of the equator; yea, ye gods! go down to the fiery pit itself, in order to keep out this frost?

But no more of this blubbering now, we are going a-whaling, and there is plenty of that yet to come. Let us scrape the ice from our frosted feet, and see what sort of a place this "Spouter" may be.

Fairhaven waterfront, circa 1880. When Melville shipped from Fairhaven in 1841, the town looked much as it does here. Union Wharf, departure point of the *Acushnet,* is on the right. Most of the houses are residences, although cooper shops, sail lofts, chandleries, taverns and other service industries are mixed in among them.

Two streets in New Bedford, circa 1870. Typical of the neighborhood of the Spouter-Inn are these streets in New Bedford. Views looking south from Union Street down First Street (opposite page) and Second Street (below) reveal a quiet, old neighborhood of rooming houses and homes of seafarers, immigrants and waterfront workers.

Millicent Library, Fairhaven

Kingman Family Collection

Chapter III

The Spouter-Inn

Sailors just landed, 1857. — *From Sherburne:* Tales of the Ocean.

ntering that gable-ended Spouter-Inn, you found yourself in a wide, low, straggling entry with old-fashioned wainscots, reminding one of the bulwarks of some condemned old craft. On one side hung a very large oil-painting so thoroughly besmoked, and every way defaced, that in the unequal cross-lights by which you viewed it, it was only by diligent study and a series of systematic visits to it, and careful inquiry of the neighbors, that you could any way arrive at an understanding of its purpose.

But what most puzzled and confounded you was a long, limber, portentous, black mass of something hovering in the centre of the picture over three blue, dim, perpendicular lines floating in a nameless yeast. A boggy, soggy, squitchy picture truly, enough to drive a nervous man distracted. Yet was there a sort of indefinite, half-attained, unimaginable sublimity about it that fairly froze you to it, till you involuntarily took an oath with yourself to find out what that marvellous painting meant. Ever and anon a bright, but, alas, deceptive idea would dart you through.—It's the Black Sea in a midnight gale.—It's the unnatural combat of the four primal elements.—It's a blasted heath.—It's a Hyperborean winter scene.—It's the breaking-up of the ice-bound stream of Time. But at last all these fancies yielded to that one portentous something in the picture's midst. *That* once found out, and all the rest were plain. But

Lower Union from Water Street, 1870s. *During Melville's day, Union Street was New Bedford's Main Street. It looked and functioned as it does in this photograph and as it still does today—a commercial district and working waterfront dominated by maritime interests, markets, taverns, ship suppliers and diners. By 1850, the five-block stretch from Purchase Street to the waterfront was said to be one of the busiest thoroughfares in all New England.* — Evening Standard, *9/27/1902. Herbert L. Aldrich photograph.*

Spinner Collection

stop; does it not bear a faint resemblance to a gigantic fish? even the great leviathan himself?

In fact, the artist's design seemed this: a final theory of my own, partly based upon the aggregated opinions of many aged persons with whom I conversed upon the subject. The picture represents a Cape-Horner in a great hurricane; the half-foundered ship weltering there with its three dismantled masts alone visible; and an exasperated whale, purposing to spring clean over the craft, is in the enormous act of impaling himself upon the three mast-heads.

The opposite wall of this entry was hung all over with a heathenish array of monstrous clubs and spears. Some were thickly set with glittering teeth resembling ivory saws; others were tufted with knots of human hair; and one was sickle-shaped, with a vast handle sweeping round like the segment made in the new-mown grass by a long-armed mower. You shuddered as you gazed, and wondered what monstrous cannibal and savage could ever have gone a death-harvesting with such a hacking, horrifying implement. Mixed with these were rusty old whaling lances and harpoons all broken and deformed. Some were storied weapons. With this once long lance, now wildly elbowed, fifty years ago did Nathan Swain kill fifteen whales between a sunrise and a sunset. And that harpoon—so like a corkscrew now—was flung in Javan seas, and run away with by a whale, years afterwards slain off the Cape of Blanco. The original iron entered nigh the tail, and, like a restless needle sojourning in the body of a man, travelled full forty feet, and at last was found imbedded in the hump.

Crossing this dusky entry, and on through yon low-arched way—cut through what in old times must have been a great central chimney with fire-places all round—you enter the public room. A still duskier place is this, with such low ponderous beams above, and such old wrinkled planks beneath, that you would almost fancy you trod some old craft's cockpits, especially of such a howling night, when this corner-anchored old ark rocked so furiously. On one side stood a long, low, shelf-like table covered with cracked glass cases, filled with dusty rarities gathered from this wide world's remotest nooks. Projecting from the further angle of the room stands a dark-looking den—the bar—a rude attempt at a right whale's head. Be that how it may, there stands the vast arched bone of the whale's jaw, so wide, a coach might almost drive beneath it. Within are shabby shelves, ranged round with old decanters, bottles, flasks; and in those jaws of swift destruction, like another cursed Jonah (by which name indeed they called him), bustles a little withered old man, who, for their money, dearly sells the sailors deliriums and death.

South Sea curios and whaling implements, 1910s. During their voyages, whalers visited exotic, often uncharted lands, collecting stories and mementos that eventually found residence in shops, taverns, museums and homes throughout New England. — *Spinner Collection.*

Upon entering the place I found a number of young seamen gathered about a table, examining by a dim light divers specimens of skrimshander. I sought the landlord, and telling him I desired to be accommodated with a room, received for answer that his house was full—not a bed unoccupied. "But avast," he added, tapping his forehead, "you hain't no objections to sharing a harpooneer's blanket, have ye? I s'pose you are goin' a whalin', so you'd better get used to that sort of thing."

I told him that I never liked to sleep two in a bed; that if I should ever do so, it would depend upon who the harpooneer might be, and that if he (the landlord) really had no other place for me, and the harpooneer was not decidedly objectionable, why rather than wander further about a strange town on so bitter a night, I would put up with the half of any decent man's blanket.

"I thought so. All right; take a seat. Supper?—you want supper? Supper'll be ready directly."

I sat down on an old wooden settle, carved all over like a bench on the Battery. At one end a ruminating tar was still further adorning it with his jack-knife, stooping over and diligently working away at the space between his legs. He was trying his hand at a ship under full sail, but he didn't make much headway, I thought.

At last some four or five of us were summoned to our meal in an adjoining room. It was cold as Iceland—no fire at all—the landlord said he couldn't afford it. Nothing but two dismal tallow candles, each in a winding sheet. We were fain to button up our monkey jackets, and hold to our lips cups of scalding tea with our half frozen fingers. But the fare was of the most substantial kind—not only meat and potatoes, but dumplings; good heavens! dumplings for supper! One young fellow in a green box coat addressed himself to these dumplings in a most direful manner.

"My boy," said the landlord, "you'll have the nightmare to a dead sartainty."

"Landlord," I whispered, "that aint the harpooneer, is it?"

"Oh, no," said he, looking a sort of diabolically funny, "the harpooneer is a dark complexioned chap. He never eats dumplings, he don't—he eats nothing but steaks, and likes em rare."

"The devil he does," says I. "Where is that harpooneer? Is he here?"

"He'll be here afore long," was the answer.

I could not help it, but I began to feel suspicious of this "dark complexioned" harpooneer. At any rate, I made up my mind that if it so turned out that we should sleep together, he must undress and get into bed before I did.

Supper over, the company went back to the bar-room, when, knowing not what else to do with myself, I resolved to spend the rest of the evening as a looker on.

Presently a rioting noise was heard without. Starting up, the landlord cried, "That's the Grampus's crew. I seed her reported in the offing this morning; a three years' voyage, and a full ship. Hurrah, boys; now we'll have the latest news from the Feegees."

A tramping of sea boots was heard in the entry; the door was flung open, and in rolled a wild set of mariners enough. Enveloped in their shaggy watch coats, and with their heads muffled in woollen comforters, all bedarned and ragged, and their beards stiff with icicles, they seemed an eruption of bears from Labrador. They had just landed from their boat, and this was the first house they entered. No wonder, then, that they made a straight wake for the whale's mouth—the bar—when the wrinkled little old Jonah, there officiating, soon poured them out brimmers all round.

The liquor soon mounted into their heads, as it generally does even with the arrantest topers newly landed from sea, and they began capering about most obstreperously.

I observed, however, that one of them held somewhat aloof, and though he seemed desirous not to spoil the hilarity of his shipmates by his own sober face, yet upon the whole he refrained from making as much noise as the rest. When the revelry of his companions had mounted to its height, this man slipped away unobserved, and I saw no more of him till he became my comrade on the sea. In a few minutes, however, he was missed by his shipmates, and being, it seems, for some reason a huge favorite with them, they raised a cry of "Bulkington! Bulkington! where's Bulkington?" and darted out of the house in pursuit of him.

It was now about nine o'clock, and the room seeming almost supernaturally quiet after these orgies, I began to congratulate myself upon a little plan that had occurred to me just previous to the entrance of the seamen.

No man prefers to sleep two in a bed. And when it comes to sleeping with an unknown stranger, in a strange inn, in a strange town, and that stranger a harpooneer, then your objections indefinitely multiply. Nor was there any earthly reason why I as a sailor should sleep two in

"Down to the Sea in Ships," 1922. Hollywood's take on the whaling industry produced some rare, authentic, high-quality footage of an American whaling voyage. Set in Quaker New Bedford of the 1840s, the film parallels settings and events created by Melville. In this scene at the "Spouter Inn," set on location in Fairhaven, whaling merchants mix it up with sailors before our hero is led down the path of intemperance, gets drunk and is shanghaied on a whaling voyage.

At Melville's Spouter Inn, Ishmael ruminates at length: "Abominable are tumblers into which he pours his poison. Thought true cylinders without—within, the villainous green goggling glasses deceitfully tapered downwards to a cheating bottom. Parallel meridians rudely pecked into the glass, surround these footpad's goblets. Fill to the mark, and your charge is but a penny; to this a penny more; and so on to the full glass—the Cape Horn measure, which you may gulp down for a shilling." — Chapter 3.

New Bedford Whaling Museum

a bed, more than anybody else; for sailors no more sleep two in a bed at sea, than bachelor Kings do ashore. To be sure they all sleep together in one apartment, but you have your own hammock, and cover yourself with your own blanket, and sleep in your own skin.

The more I pondered over this harpooneer, the more I abominated the thought of sleeping with him. Suppose now, he should tumble in upon me at midnight—how could I tell from what vile hole he had been coming?

"Landlord! I've changed my mind about that harpooneer.—I shan't sleep with him. I'll try the bench here."

"Just as you please; I'm sorry I can't spare ye a tablecloth for a mattress, and it's a plaguy rough board here"—feeling of the knots and notches. "But wait a bit, Skrimshander; I've got a carpenter's plane there in the bar—wait, I say, and I'll make ye snug enough." So saying he procured the plane; and with his old silk handkerchief first dusting the bench, vigorously set to planing away at my bed, the while grinning like an ape. The shavings flew right and left; till at last the plane-iron came bump against an indestructible knot. The landlord was near spraining his wrist, and I told him for heaven's sake to quit—the bed was soft enough to suit me, and I did not know how all the planing in the world could make eider down of a pine plank. So gathering up the shavings with another grin, and throwing them into the great stove in the middle of the room, he went about his business, and left me in a brown study.

I now took the measure of the bench, and found that it was a foot too short; but that could be mended with a chair. But it was a foot too narrow, and the other bench in the room was about four inches higher than the planed one—so there was no yoking them. I then placed the first bench lengthwise along the only clear space against the wall, leaving a little interval between, for my back to settle down in. But I soon found that there came such a draught of cold air over me from under the sill of the window, that this plan would never do at all, especially as another current from the rickety door met the one from the window, and both together formed a series of small whirlwinds in the immediate vicinity of the spot where I had thought to spend the night.

The devil fetch that harpooneer, thought I, but stop, couldn't I steal a march on him—bolt his door inside, and jump into his bed, not to be wakened by the most violent knockings? It seemed no bad idea; but upon second thoughts I dismissed it. For who could tell but what the next morning, so soon as I popped out of the room, the harpooneer might be standing in the entry, all ready to knock me down!

Still, looking around me again, and seeing no possible chance of spending a sufferable night unless in some

Mariners' Home, 1880. Next door to the Seaman's Bethel is the Mariners' Home, where seafarers have found temporary shelter at minimal cost for 150 years. Built about 1787 as the home of future whaling magnate William Rotch, Jr., it was donated to the New Bedford Port Society in 1851 by his daughter, Sarah Rotch Arnold. — Kingman Family Collection.

other person's bed, I began to think that after all I might be cherishing unwarrantable prejudices against this unknown harpooneer. Thinks I, I'll wait awhile; he must be dropping in before long. I'll have a good look at him then, and perhaps we may become jolly good bedfellows after all—there's no telling.

But though the other boarders kept coming in by ones, twos, and threes, and going to bed, yet no sign of my harpooneer.

"Landlord!" said I, "what sort of a chap is he—does he always keep such late hours?" It was now hard upon twelve o'clock.

The landlord chuckled again with his lean chuckle, and seemed to be mightily tickled at something beyond my comprehension. "No," he answered, "generally he's an early bird—airley to bed and airley to rise—yea, he's the bird what catches the worm.—But to-night he went out a peddling, you see, and I don't see what on airth keeps him so late, unless, may be, he can't sell his head."

"Can't sell his head?—What sort of a bamboozingly story is this you are telling me?" getting into a towering rage. "Do you pretend to say, landlord, that this harpooner is actually engaged this blessed Saturday night, or rather Sunday morning, in peddling his head around this town?"

"That's precisely it," said the landlord, "and I told him he couldn't sell it here, the market's overstocked."

"With what?" shouted I.

"With heads to be sure; aint there too many heads in the world?"

"I tell you what it is, landlord," said I, quite calmly, "you'd better stop spinning that yarn to me—I'm not green."

"May be not," taking out a stick and whittling a toothpick, "but I rayther guess you'll be done brown if that 'ere harpooneer hears you a slanderin' his head."

"I'll break it for him," said I, now flying into a passion again at this unaccountable farrago of the landlord's.

"It's broke a'ready," said he.

"Broke," said I—"*broke*, do you mean?"

"Sartain, and that's the very reason he can't sell it, I guess."

"Landlord," said I, going up to him as cool as Mt. Hecla in a snow storm,—"landlord, stop whittling. You and I must understand one another, and that too without delay. I now demand of you to speak out and tell me who and what this harpooneer is, and whether I shall be in all respects safe to spend the night with him. And in the first place, you will be so good as to unsay that story about selling his head, which if true I take to be good evidence that this harpooneer is stark mad, and I've no idea of sleeping with a madman; and you, sir, *you* I mean, landlord, *you*, sir, by trying to induce me to do so knowing'y, would thereby render yourself liable to a criminal prosecution."

"Wall," said the landlord, fetching a long breath, "that's a purty long sarmon for a chap that rips a little now and then. But be easy, be easy, this here harpooneer I have been tellin' you of has just arrived from the south seas, where he bought up a lot of 'balmed New Zealand heads (great curios, you know), and he's sold all on 'em but one, and that one he's trying to sell to-night, cause to-morrow's Sunday, and it would not do to be sellin' human heads about the streets when folks is goin' to churches. He wanted to, last Sunday, but I stopped him just as he was goin' out of the door with four heads strung on a string, for all the airth like a string of inions."

This account cleared up the otherwise unaccountable mystery, and showed that the landlord, after all, had had no idea of fooling me—but at the same time what could I think of a harpooneer who stayed out of a Saturday night clean into the holy Sabbath, engaged in such a cannibal business as selling the heads of dead idolators?

"Depend upon it, landlord, that harpooneer is a dangerous man."

"He pays reg'lar," was the rejoinder. "But come, it's getting dreadful late, you had better be turning flukes—it's a nice bed: Sal and me slept in that 'ere bed the night we were spliced. There's plenty room for two to kick about in that bed; it's an almighty big bed that. Come along here, I'll give ye a glim in a jiffy;" and so saying he lighted a candle and held it towards me, offering to lead the way. But I stood irresolute; when looking at a clock in the corner, he exclaimed "I vum it's Sunday—you won't see that harpooneer to-night; he's come to anchor somewhere—come along then; *do* come; *won't ye come?*"

I considered the matter a moment, and then up stairs we went, and I was ushered into a small room, cold as a clam, and furnished, sure enough, with a prodigious bed, almost big enough indeed for any four harpooneers to sleep abreast.

"There," said the landlord, placing the candle on a crazy old sea chest that did double duty as a wash-stand and centre table; "there, make yourself comfortable now, and good night to ye." I turned round from eyeing the bed, but he had disappeared.

Folding back the counterpane, I stooped over the bed. Though none of the most elegant, it yet stood the scrutiny tolerably well. I then glanced round the room; and besides the bedstead and centre table, could see no other furniture belonging to the place, but a rude shelf, the four walls, and a papered fireboard representing a man striking a whale. Of things not properly belonging to the room, there was a hammock lashed up, and thrown upon the floor in one corner; also a large seaman's bag, containing the harpooneer's wardrobe, no doubt in lieu of a land trunk. Likewise, there was a parcel of outlandish bone fish hooks on the shelf over the fireplace, and a tall harpoon standing at the head of the bed.

But what is this on the chest? I can compare it to nothing but a large door mat, ornamented at the edges with little tinkling tags something like the stained porcupine quills round an Indian moccasin.

I sat down on the side of the bed, and commenced thinking about this head-peddling harpooneer, and his door mat. After thinking some time on the bed-side, I got up and took off my monkey jacket, and then stood in the middle of the room thinking. I then took off my coat, and thought a little more in my shirt sleeves. But beginning to feel very cold now, half undressed as I was, and remembering what the landlord said about the harpooneer's not coming home at all that night, it being so very late, I made no more ado, but jumped out of my pantaloons and boots, and then blowing out the light tumbled into bed, and commended myself to the care of heaven.

Whether that mattress was stuffed with corn-cobs or broken crockery, I rolled about a good deal, and could not sleep for a long time. At last I slid off into a light doze, and had pretty nearly made a good offing towards the land of Nod, when I heard a heavy footfall in the passage, and saw a glimmer of light come into the room from under the door.

Lord save me, thinks I, that must be the harpooner, the infernal head-peddlar. I lay perfectly still, and resolved not to say a word till spoken to. Holding a light in one hand, and that identical New Zealand head in the other, the stranger entered the room, and without looking towards the bed, placed his candle a good way off from me on the floor in one corner, and then began working away at the knotted cords of the large bag I before spoke of as being in the room. I was all eagerness to see his face, but he kept it averted for some time while employed in unlacing the bag's mouth. This accomplished, however, he turned round—when, good Heavens! What a sight! Such a face! It was of a dark purplish, yellow color, here and there stuck over with large, blackish looking squares. Yes, it's just as I thought, he's a terrible bedfellow: he's been in a fight, got dreadfully cut, and here he is, just from the surgeon. But at that moment he chanced to turn his face so towards the light, that I plainly saw they could not be sticking-plasters at all, those black squares on his cheeks. They were stains of some sort or other. At first I knew not what to make of this; but soon an inkling of the truth occurred to me. I remembered a story of a white man—a whaleman too—who, falling among the cannibals, had been tattooed by them. I concluded that this harpooneer, in the course of his distant voyages, must have met with a similar adventure. And what is it, thought I, after all! It's only his outside; a man can be honest in any sort of skin. Now, while all these ideas were passing through me like lightning, this harpooneer never noticed me at all. But, after some difficulty having opened his bag, he commenced fumbling in it, and presently pulled out a sort of tomahawk, and a seal-skin wallet with the hair on. Placing these on the old chest in the middle of the room, he then took the New Zealand head—a ghastly thing enough—and crammed it down into the bag. He now took off his hat—a new beaver hat. There was no hair on his head—none to speak of at least—nothing but a small scalp-knot twisted up on his forehead. His bald purplish head now looked for all the world like a mildewed skull. Had not the stranger stood between me and the door, I would have bolted out of it quicker than ever I bolted a dinner.

Even as it was, I thought something of slipping out of the window, but it was the second floor back. I am no coward, but what to make of this head-peddling purple rascal altogether passed my comprehension. Ignorance is the parent of fear, and being completely nonplussed and confounded about the stranger, I confess I was now as much afraid of him as if it was the devil himself who had thus broken into my room at the dead of night.

Meanwhile, he continued the business of undressing, and at last showed his chest and arms. As I live, these covered parts of him were checkered with the same squares as his face; his back, too, was all over the same dark squares. Still more, his very legs were marked, as if a parcel of dark green frogs were running up the trunks of young palms. It was now quite plain that he must be some abominable savage or other shipped aboard of a whaleman in the South Seas, and so landed in this Christian country. I quaked to think of it. A peddler of heads too—perhaps the heads of his own brothers. He might take a fancy to mine—heavens! look at that tomahawk!

But there was no time for shuddering, for now the savage went about something that completely fascinated my attention, and convinced me that he must indeed be a heathen. Going to his heavy grego, or wrapall, or dreadnaught, which he had previously hung on a chair, he fumbled in the pockets, and produced at length a curious little deformed image with a hunch on its back, and exactly the color of a three days' old Congo baby. Remembering the embalmed head, at first I almost thought that this black manikin was a real baby preserved in some similar manner. But seeing that it was not at all limber, and that it glistened a good deal like polished ebony, I concluded that it must be nothing but a wooden idol, which indeed it proved to be. For now the

Maori chiefs. These engravings depict a New Zealand chief (right) and Otegoowgoow, the son of a chief—their faces "curiously tattooed or marked according to their manner." It is widely believed that Queequeg and his native Rokovoko are fashioned from the Maori of New Zealand, who are thought to have migrated in early times from other Polynesian islands. Maori tradition asserts their ancestors came to New Zealand in seven canoes. The Maori language is closely related to Rarotongan, Tahitian and other languages spoken in the South Pacific. The tattoo, based on the Polynesian term "tatu," is developed to its highest level among the Maori and in the Marquesas, where high-status men were completely tattooed.

Kendall Collection, NBWM

Maori carving, 1836. Besides tattooing, the Maori were skilled in the art of carving. Although this figure appears at the top of a post, similar human figures with crossed legs were carved from a soft green stone and worn around the neck. Like Queequeg's Yojo, these "leitiki" were considered sacred and were usually passed down from one generation to the next. — *Illustration by Alfred T. Agate. From Wilkes:* Narrative of the United States.... *New Bedford Whaling Museum.*

savage goes up to the empty fire-place, and removing the papered fire-board, sets up this little hunchbacked image, like a tenpin, between the andirons. The chimney jambs and all the bricks inside were very sooty, so that I thought this fire-place made a very appropriate little shrine or chapel for his Congo idol.

I now screwed my eyes hard towards the half hidden image, feeling but ill at ease meantime—to see what was next to follow. First he takes about a double handful of shavings out of his grego pocket, and places them carefully before the idol; then laying a bit of ship biscuit on top and applying the flame from the lamp, he kindled the shavings into a sacrificial blaze. Presently, after many hasty snatches into the fire, and still hastier withdrawals of his fingers (whereby he seemed to be scorching them badly), he at last succeeded in drawing out the biscuit; then blowing off the heat and ashes a little, he made a polite offer of it to the little negro. But the little devil did not seem to fancy such dry sort of fare at all; he never moved his lips. All these strange antics were accompanied by still stranger guttural noises from the devotee, who seemed to be praying in a sing-song or else singing some pagan psalmody or other, during which his face twitched about in the most unnatural manner. At last extinguishing the fire, he took the idol up very unceremoniously, and bagged it again in his grego pocket as carelessly as if he were a sportsman bagging a dead woodcock.

All these queer proceedings increased my uncomfortableness, and seeing him now exhibiting strong symptoms of concluding his business operations, and jumping into bed with me, I thought it was high time to break the spell into which I had so long been bound.

But the interval I spent in deliberating what to say, was a fatal one. Taking up his tomahawk from the table, he examined the head of it for an instant, and then holding it to the light, with his mouth at the handle, he puffed out great clouds of tobacco smoke. The next moment the light was extinguished, and this wild cannibal, tomahawk between his teeth, sprang into bed with me. I sang out, I could not help it now; and giving a sudden grunt of astonishment he began feeling me.

Stammering out something, I knew not what, I rolled away from him against the wall, and then conjured him, whoever or whatever he might be, to keep quiet, and let me get up and light the lamp again. But his guttural responses satisfied me at once that he but ill comprehended my meaning.

"Who-e debel you?"—he at last said—"you no speak-e, dam-me, I kill-e." And so saying the lighted tomahawk began flourishing about me in the dark.

"Landlord, for God's sake, Peter Coffin!" shouted I. "Landlord! Watch! Coffin! Angels! save me!"

"Speak-e! tell-ee me who-ee be, or dam-me, I kill-e!" again growled the cannibal, while his horrid flourishings of the tomahawk scattered the hot tobacco ashes about me till I thought my linen would get on fire. But thank heaven, at that moment the landlord came into the room light in hand, and leaping from the bed I ran up to him.

"Don't be afraid now," said he, grinning again. "Queequeg here wouldn't harm a hair of your head."

"Stop your grinning," shouted I, "and why didn't you tell me that that infernal harpooneer was a cannibal?"

"I thought ye know'd it;—didn't I tell ye, he was peddlin' heads around town?—but turn flukes again and go to sleep. Queequeg, look here—you sabbee me, I sabbee you—this man sleepe you—you sabbee?"—

"Me sabbee plenty"—grunted Queequeg, puffing away at his pipe and sitting up in bed.

"You gettee in," he added, motioning to me with his tomahawk, and throwing the clothes to one side. He really did this in not only a civil but a really kind and charitable way. I stood looking at him a moment. For all his tattooings he was on the whole a clean, comely looking cannibal. Better sleep with a sober cannibal than a drunken Christian.

"Landlord," said I, "tell him to stash his tomahawk there, or pipe, or whatever you call it; tell him to stop smoking, in short, and I will turn in with him. But I don't fancy having a man smoking in bed with me. It's dangerous. Besides, I aint insured."

This being told to Queequeg, he at once complied, and again politely motioned me to get into bed—rolling over to one side as much as to say—I won't touch a leg of ye.

"Good night, landlord," said I, "you may go."

I turned in, and never slept better in my life.

Chapter IV

The Counterpane

"Ishmael and Queequeg." — *Illustration by Rockwell Kent.*
©1930 by R. R. Donnelley & Sons Co., with permission.

Upon waking next morning about daylight, I found Queequeg's arm thrown over me in the most loving and affectionate manner. The counterpane was of patchwork, full of odd little parti-colored squares and triangles; and this arm of his tattooed all over with an interminable Cretan labyrinth of a figure, no two parts of which were of one precise shade—owing I suppose to his keeping his arm at sea unmethodically in sun and shade, his shirt sleeves irregularly rolled up at various times—looked for all the world like a strip of that same patchwork quilt. It was only by the sense of weight and pressure that I could tell that Queequeg was hugging me.

I tried to move his arm—unlock his bridegroom clasp—yet, sleeping as he was, he still hugged me tightly, as though naught but death should part us twain. I now strove to rouse him—"Queequeg!"—but his only answer was a snore. I then rolled over, my neck feeling as if it were in a horse-collar; and suddenly felt a slight scratch. Throwing aside the counterpane, there lay the tomahawk sleeping by the savage's side, as if it were a hatchet-faced baby. "Queequeg!—in the name of goodness, Queequeg, wake!" At length, by dint of much wriggling, and loud and incessant expostulations upon the unbecomingness of his hugging a fellow male in that matrimonial sort of style, I succeeded in extracting a grunt; and presently, he drew back his arm, shook himself all over like a Newfoundland dog just from the water, and sat up in bed, stiff as a pike-staff, looking at me, and rubbing his eyes as if he did not altogether remember how I came to be there. Meanwhile, I lay quietly eyeing him, having no serious misgivings now, and bent upon narrowly observing so curious a creature. When, at last, his mind seemed made up touching the character of his bedfellow, and he became, as it were, reconciled to the fact; he jumped out upon the floor, and by certain signs and sounds gave me to understand that, if it pleased me, he would dress first and then leave me to dress afterwards, leaving the whole apartment to myself. Thinks I, Queequeg, under the circumstances, this is a very civilized overture; but, the truth is these savages have an innate sense of delicacy, say what you will; it is marvellous how essentially polite they are.

He commenced dressing at top by donning his beaver hat, a very tall one, by the by, and then—still minus his trowsers—he hunted up his boots. What under the heavens he did it for, I cannot tell, but his next movement was to crush himself—boots in hand, and hat on—under the bed; when, from sundry violent gaspings and strainings, I inferred he was hard at work booting himself; though by no law of propriety that I ever heard of, is any man required to be private when putting on his boots. But Queequeg, do you see, was a creature in the transition state—neither caterpillar nor butterfly. He was just enough civilized to show off his outlandishness in the strangest possible manner. At last, he emerged with his hat very much dented and crushed down over his eyes, and began creaking and limping about the room, as if, not being much accustomed to boots, his pair of damp, wrinkled cowhide ones—probably not made to order either—rather pinched and tormented him at the first go off of a bitter cold morning.

Seeing, now, that there were no curtains to the window, and that the street being very narrow, the house opposite commanded a plain view into the room, and observing more and more the indecorous figure that Queequeg made, staving about with little else but his hat and boots on; I begged him as well as I could, to accelerate his toilet somewhat, and particularly to get into his pantaloons as soon as possible. He complied, and then proceeded to wash himself. At that time in the morning any Christian would have washed his face; but Queequeg, to my amazement, contented himself with restricting his ablutions to his chest, arms, and hands. He then donned his waistcoat, and taking up a piece of hard soap on the wash-stand centre-table, dipped it into water and commenced lathering his face. I was watching to see where he kept his razor, when lo and behold, he takes the harpoon from the bed corner, slips out the long wooden stock, unsheathes the head, whets it a little on his boot, and striding up to the bit of mirror against the wall, begins a vigorous scraping, or rather harpooning of his cheeks. Afterwards I wondered the less at this operation when I came to know of what fine steel the head of a harpoon is made, and how exceedingly sharp the long straight edges are always kept.

The rest of his toilet was soon achieved, and he proudly marched out of the room, wrapped up in his great pilot monkey jacket, and sporting his harpoon like a marshal's baton.

Chapter V, VI

Breakfast · The Street

Union Street, 1865. — *Kingman Family Collection.*

quickly followed suit, and descending into the bar-room accosted the grinning landlord very pleasantly. I cherished no malice towards him, though he had been skylarking with me not a little in the matter of my bedfellow.

The bar-room was now full of the boarders who had been dropping in the night previous. They were nearly all whalemen; chief mates, and second mates, and third mates, and sea carpenters, and sea coopers, and sea blacksmiths, and harpooneers, and ship keepers; a brown and brawny company, with bosky beards; an unshorn, shaggy set, all wearing monkey jackets for morning gowns.

You could pretty plainly tell how long each one had been ashore. This young fellow's healthy cheek is like a sun-toasted pear in hue, and would seem to smell almost as musky; he cannot have been three days landed from his Indian voyage. That man next him looks a few shades lighter; you might say a touch of satin wood is in him. In the complexion of a third still lingers a tropic tawn, but slightly bleached withal; *he* doubtless has tarried whole weeks ashore.

"Grub, ho!" now cried the landlord, flinging open a door, and in we went to breakfast.

They say that men who have seen the world, thereby become quite at ease in manner, quite self-possessed in company. Not always, though: Ledyard, the great New England traveller, and Mungo Park, the Scotch one; of all men, they possessed the least assurance in the parlor. This kind of travel, I say, may not be the very best mode of attaining a high social polish.

After we were all seated at the table, I was preparing to hear some good stories about whaling; to my no small surprise, nearly every man maintained a profound silence. And not only that, but they looked embarrassed. Yea, here were a set of sea-dogs, many of whom without the slightest bashfulness had boarded great whales on the high seas—entire strangers to them—and duelled

Water Street, 1880s. *Throughout the last half of the 19th century, Water Street was the Mecca for the pilgrim mariner. Here, the whaleman's every need could be met. This row between William and Union streets features "Rooms to Let—Washing and Ironing," a meat market, a ship store, dining rooms, "Tower's Fish Brand—Oiled Whiting," Thomas Donaghy's "Oil and Rubber Goods," and the Foreign Emigration Office.*

Kingman Family Collection

them dead without winking; and yet, here they sat at a social breakfast table, looking round as sheepishly at each other as though they had never been out of sight of some sheepfold among the Green Mountains.

But as for Queequeg—why, Queequeg sat there among them—at the head of the table, too, it so chanced; as cool as an icicle. To be sure I cannot say much for his breeding. His greatest admirer could not have cordially justified his bringing his harpoon into breakfast with him, and using it there without ceremony; reaching over the table with it, to the imminent jeopardy of many heads, and grappling the beefsteaks towards him. But *that* was certainly very coolly done by him, and every one knows that in most people's estimation, to do anything coolly is to do it genteelly.

We will not speak of all Queequeg's peculiarities here; how he eschewed coffee and hot rolls, and applied his undivided attention to beefsteaks, done rare. Enough, that when breakfast was over he withdrew like the rest into the public room, lighted his tomahawk-pipe, and was sitting there quietly digesting and smoking with his inseparable hat on, when I sallied out for a stroll.

CHAPTER VI

f I had been astonished at first catching a glimpse of so outlandish an individual as Queequeg circulating among the polite society of a civilized town, that astonishment soon departed upon taking my first daylight stroll through the streets of New Bedford.

In thoroughfares nigh the docks, any considerable seaport will frequently offer to view the queerest looking nondescripts from foreign parts. Even in Broadway and

Try pots and blubber hooks, 1910s. Discarded remnants of the whaling trade were a common sight on the streets of New Bedford, even well into the 20th century. — *Joseph S. Martin photograph. William R. Hegarty Collection.*

Chestnut streets, Mediterranean mariners will sometimes jostle the affrighted ladies. Regent street is not unknown to Lascars and Malays; and at Bombay, in the Apollo Green, live Yankees have often scared the natives. But New Bedford beats all Water street and Wapping. In these last-mentioned haunts you see only sailors; but in New Bedford, actual cannibals stand chatting at street corners; savages outright; many of whom yet carry on their bones unholy flesh. It makes a stranger stare.

But, besides the Feegeeans, Tongatabooarrs, Erromanggoans, Pannangians, and Brighggians, and, besides the wild specimens of the whaling-craft which unheeded reel about the streets, you will see other sights still more curious, certainly more comical. There weekly arrive in this town scores of green Vermonters and New Hampshire men, all athirst for gain and glory in the fishery. They are mostly young, of stalwart frames; fellows who have felled forests, and now seek to drop the axe and snatch the whale-lance. In some things you would think them but a few hours old. Look there! that chap strutting round the corner. He wears a beaver hat and swallow-tailed coat, girdled with a sailor-belt and

Water Street, 1870s. Merchants on Water Street provided an eclectic mix of fare and consumer goods; and an assortment of banks, brokerage firms, counting houses and other financial enterprises offered their services. Here, sailors, merchants and tycoons intermingled.

"…In New Bedford, actual cannibals stand chatting at street corners…. It makes a stranger stare. But, besides the wild specimens of the whaling-craft which unheeded reel about the streets, you will see other sights still more curious, certainly more comical." — *Chapter 6.*

On this corner (with William Street), you can shop at C. R. Sherman & Co., dealers in sextants and other nautical instruments, or bank at one of three banks. Across the street, signs on buildings indicate that blasting powder and fish are for sale.

Kingman Family Collection

sheath-knife. Here comes another with a sou'-wester and a bombazine cloak.

No town-bred dandy will compare with a country-bred one—I mean a downright bumpkin dandy—a fellow that, in the dog-days, will mow his two acres in buckskin gloves for fear of tanning his hands. Now when a country dandy like this takes it into his head to make a distinguished reputation, and joins the great whale-fishery, you should see the comical things he does upon reaching the seaport. In bespeaking his sea-outfit, he orders bell-buttons to his waistcoats; straps to his canvas trowsers. Ah, poor Hay-Seed! how bitterly will burst those straps in the first howling gale, when thou art driven, straps, buttons, and all, down the throat of the tempest.

But think not that this famous town has only harpooneers, cannibals, and bumpkins to show her visitors. Not at all. Still New Bedford is a queer place. Had it not been for us whalemen, that tract of land would this day perhaps have been in as howling condition as the coast of Labrador. As it is, parts of her back country are enough to frighten one, they look so bony. The town itself is perhaps the dearest place to live in, in all New England. It is a land of oil, true enough; but not like Canaan; a land, also, of corn and wine. The streets do not run with milk; nor in the spring-time do they pave them with fresh eggs. Yet, in spite of this, nowhere in all America will you find more patrician-like houses; parks and gardens more opulent, than in New Bedford. Whence came they? how planted upon this once scraggy scoria of a country?

Go and gaze upon the iron emblematical harpoons round yonder lofty mansion, and your question will be answered. Yes; all these brave houses and flowery gar-

Union Street, 1860s. *In Melville's day (early 1840s), the four-block stretch along Union Street to the waterfront was said to be one of the busiest thoroughfares in all New England, housing 150 businesses including artists, tailors, physicians, architects, hostelries and dining rooms.* — Evening Standard, 9/27/1902. Kingman Family Collection.

dens came from the Atlantic, Pacific, and Indian oceans. One and all, they were harpooned and dragged up hither from the bottom of the sea.

In New Bedford, fathers, they say, give whales for dowers to their daughters, and portion off their nieces with a few porpoises a-piece. You must go to New Bedford to see a brilliant wedding; for, they say, they have reservoirs of oil in every house, and every night recklessly burn their lengths in spermaceti candles.

The patrician-like Parker home, County Street, 1880. *New Bedford's first millionaire, John Avery Parker, commissioned renowned Rhode Island architect Russell Warren to design this home in 1833. It was said to be the largest Greek Revival residence ever built in the United States.*

Kingman Family Collection

County Street, 1860s. *New Bedford's grandest avenue was, and still is, County Street. Once noted for its stately elms, grand oaks and numerous congregations, the street was developed primarily by whaling merchants. This view looks south from the front of Charles W. Morgan's estate at the head of William Street.* — Spinner Collection.

James Arnold House, County Street, 1870s. *The home of whaling merchant James Arnold, later purchased by William Rotch, Sr., was famous for its magnificent gardens. Melville is known to have visited these gardens on his New Bedford visits. James Arnold was an avid horticulturist who cultivated hundreds of acres of gardens, trees and other flora around his estate. Arnold donated his botanical legacy to Harvard University, where it became the world-renowned Arnold Arboretum.*

In summer time, the town is sweet to see; full of fine maples—long avenues of green and gold. And in August, high in air, the beautiful and bountiful horse-chestnuts, candelabra-wise, proffer the passer-by their tapering upright cones of congregated blossoms. So omnipotent is art; which in many a district of New Bedford has superinduced bright terraces of flowers upon the barren refuse rocks thrown aside at creation's final day.

And the women of New Bedford, they bloom like their own red roses. But roses only bloom in summer; whereas the fine carnation of their cheeks is perennial as sunlight in the seventh heavens. Elsewhere match that bloom of theirs, ye cannot, save in Salem, where they tell me the young girls breathe such musk, their sailor sweethearts smell them miles off shore, as though they were drawing nigh the odorous Moluccas instead of the Puritanic sands.

The Rotches. *New Bedford's whaling "Rockerfellers" were the Rotches, led by Joseph Rotch who emigrated from Nantucket in 1767 and purchased land with the intent of establishing his enterprise, replete with ships, factories, suppliers and agents. His son William further advanced the business and his grandson, William, Jr. (right), presided over the world's largest whaling fleet when the industry reached its apex in the 1840s. Perhaps Melville had William's cousin Elizabeth in mind (below) when he mused about the delicate beauty of New Bedford women: "the fine carnation of their cheeks is perennial as sunlight in the seventh heavens."* — Wm. Rotch, Jr. portrait by William A. Wall, ca. 1835, private collection. Elizabeth Rotch portrait by Thomas Sully, 1833, New Bedford Whaling Museum.

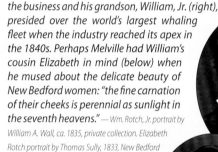

New Bedford Whaling Museum

Chapter VII, VIII

The Chapel · The Pulpit

In this same New Bedford there stands a Whaleman's Chapel, and few are the moody fishermen, shortly bound for the Indian Ocean or Pacific, who fail to make a Sunday visit to the spot. I am sure that I did not.

Returning from my first morning stroll, I again sallied out upon this special errand. The sky had changed from clear, sunny cold, to driving sleet and mist. Wrapping myself in my shaggy jacket of the cloth called bearskin, I fought my way against the stubborn storm. Entering, I found a small scattered congregation of sailors, and sailors' wives and widows. A muffled silence reigned, only broken at times by the shrieks of the storm. Each silent worshipper seemed purposely sitting apart from the other, as if each silent grief were insular and incommunicable. The chaplain had not yet arrived; and there these silent islands of men and women sat steadfastly eyeing several marble tablets, with black borders, masoned into the wall on either side the pulpit. Three of them ran something like the following, but I do not pretend to quote:—

SACRED
To the Memory
OF
JOHN TALBOT,
Who, at the age of eighteen, was lost overboard,
Near the Isle of Desolation, off Patagonia,
November 1st, 1836.
THIS TABLET
Is erected to his Memory
BY
BY HIS SISTER.

———

SACRED
To the Memory
OF
The late
CAPTAIN EZEKIEL HARDY,
Who in the bows of his boat was killed by a
Sperm Whale on the coast of Japan,
August 3rd, 1833.
THIS TABLET
Is erected to his Memory
BY
HIS WIDOW.

———

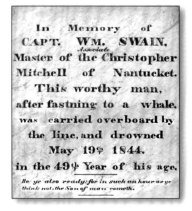

Bethel tablet. — *Seth Beall photograph.*

Shaking off the sleet from my ice-glazed hat and jacket, I seated myself near the door, and turning sideways was surprised to see Queequeg near me. Affected by the solemnity of the scene, there was a wondering gaze of incredulous curiosity in his countenance. This savage was the only person present who seemed to notice my entrance; because he was the only one who could not read, and, therefore, was not reading those frigid inscriptions on the wall. Whether any of the relatives of the seamen whose names appeared there were now among the congregation, I knew not; but so many are the unrecorded accidents in the fishery, and so plainly did several women present wear the countenance if not the trappings of some unceasing grief, that I feel sure that here before me were assembled those, in whose unhealing hearts the sight of those bleak tablets sympathetically caused the old wounds to bleed afresh.

Oh! ye whose dead lie buried beneath the green grass; who standing among flowers can say—here, *here* lies my beloved; ye know not the desolation that broods in bosoms like these. What bitter blanks in those black-bordered marbles which cover no ashes! What despair in those immovable inscriptions! What deadly voids and unbidden infidelities in the lines that seem to gnaw upon all Faith, and refuse resurrections to the beings who have placelessly perished without a grave.

In what census of living creatures, the dead of mankind are included; why it is that a universal proverb says of them, that they tell no tales, though containing more secrets than the Goodwin Sands; how it is that to his name who yesterday departed for the other world, we prefix so significant and infidel a word, and yet do not thus entitle him, if he but embarks for the remotest Indies of this living earth; why the Life Insurance Companies pay death- forfeitures upon immortals; in what eternal, unstirring paralysis, and deadly, hopeless trance, yet lies antique Adam who died sixty round centuries ago; how it is that we still refuse to be comforted for those who we nevertheless maintain are dwelling in unspeakable bliss; why all the living so strive to hush all the dead; wherefore but the rumor of a knocking in a tomb will terrify a whole city. All these things are not without their meanings.

But Faith, like a jackal, feeds among the tombs, and even from these dead doubts she gathers her most vital hope.

It needs scarcely to be told, with what feelings, on the eve of a Nantucket voyage, I regarded those marble tablets, and by the murky light of that darkened, doleful day read the fate of the whalemen who had gone before me. Yes, Ishmael, the same fate may be thine. But somehow I grew merry again. Delightful inducements to embark, fine chance for promotion, it seems—aye, a stove boat will make me an immortal by brevet. Yes, there is death in this business of whaling—a speechlessly quick chaotic bundling of a man into Eternity. But what then? Methinks we have hugely mistaken this matter of Life and Death. Methinks that what they call my shadow here on earth is my true substance. Methinks that in looking at things spiritual, we are too much like oysters observing the sun through the water, and thinking that thick water the thinnest of air. Methinks my body is but the lees of my better being. In fact take my body who will, take it I say, it is not me. And therefore three cheers for Nantucket; and come a stove boat and stove body when they will, for stave my soul, Jove himself cannot.

CHAPTER VII

I had not been seated very long ere a man of certain venerable robustness entered; immediately as the storm-pelted door flew back upon admitting him, a quick regardful eyeing of him by all the congregation, sufficiently attested that this fine old man was the chaplain. Yes, it was the famous Father Mapple, so called by the whalemen, among whom he was a great favorite. He had been a sailor and a harpooneer in his youth, but for many years past had dedicated his life to the ministry. Hat and coat and overshoes were one by one removed; when arrayed in a decent suit, he quietly approached the pulpit.

Like most old fashioned pulpits, it was a very lofty one, and the architect, it seemed, had acted upon the hint of Father Mapple, and finished the pulpit without a stairs, substituting a perpendicular side ladder, like those used in mounting a ship from a boat at sea. The perpendicular parts of this side ladder were of cloth-covered rope, so that at every step there was a joint. I was not prepared to see Father Mapple after gaining the height, slowly turn round, and stooping over the pulpit, deliberately drag up the ladder step by step, till the whole was deposited within, leaving him impregnable in his little Quebec.

The Bethel walls honor those who went down to the sea in ships. — *Seth Beall photograph.*

But the side ladder was not the only strange feature of the place, borrowed from the chaplain's former sea-farings. Between the marble cenotaphs on either hand of the pulpit, the wall which formed its back was adorned with a large painting representing a gallant ship beating against a terrible storm off a lee coast of black rocks and snowy breakers. But high above the flying scud and dark-rolling clouds, there floated a little isle of sunlight, from which beamed forth an angel's face; and this bright face shed a distant spot of radiance upon the ship's tossed deck.

Nor was the pulpit itself without a trace of the same sea-taste that had achieved the ladder and the picture. Its paneled front was in the likeness of a ship's bluff bows, and the Holy Bible rested on a projecting piece of scroll work, fashioned after a ship's fiddle-headed beak.

"The Pulpit," 1954. *John Huston's version of the pulpit, installed at the Bethel for his film classic, "Moby Dick," is a credible representation of Melville's pulpit: "Nor was the pulpit itself without a trace of the same sea-taste that had achieved the ladder and the picture. Its panelled front was in the likeness of a ship's bluff bows, and the Holy Bible rested on the projecting piece of scroll work, fashioned after a ship's fiddle-headed beak." Still vital to seafarers and community members, the Bethel, operated by the New Bedford Port Society, is open for services—its walls still marked with epithets of vanquished seamen from Melville's time to the present day.*

Courtesy of Paul Levasseur

Chapter ix

The Sermon

Father Mapple rose, and in a mild voice of unassuming authority ordered the scattered people to condense. "Starboard gangway, there! side away to larboard—larboard gangway to starboard! Midships! midships!"

There was a low rumbling of heavy sea-boots among the benches, and a still slighter shuffling of women's shoes, and all was quiet again, and every eye on the preacher.

A brief pause ensued; the preacher slowly turned over the leaves of the Bible, and at last, folding his hand down upon the proper page, said: "Beloved shipmates, clinch the last verse of the first chapter of Jonah—'And God had prepared a great fish to swallow up Jonah.'

"What is this lesson that the book of Jonah teaches? Shipmates, it is a two-stranded lesson; a lesson to us all as sinful men, and a lesson to me as a pilot of the living God. As with all sinners among men, the sin was in his wilful disobedience of the command of God. And if we obey God, we must disobey ourselves; and it is in this disobeying ourselves, wherein the hardness of obeying God consists.

"With this sin of disobedience in him, Jonah still further flouts at God, by seeking to flee from Him. He thinks that a ship made by men, will carry him into countries where God does not reign, but only the Captains of this earth. He skulks about the wharves of Joppa, and seeks a ship that's bound for Tarshish.

"Now Jonah's Captain, shipmates, was one whose discernment detects crime in any, but whose cupidity exposes it only in the penniless. In this world, shipmates, sin that pays its way can travel freely, and without a passport; whereas Virtue, if a pauper, is stopped at all frontiers. So Jonah's Captain prepares to test the length of Jonah's purse, ere he judge him openly. He charges him thrice the usual sum; and it's assented to. Then the Captain knows that Jonah is a fugitive; but at the same time resolves to help a flight that paves its rear with gold. Yet when Jonah fairly takes out his purse, prudent suspicions still molest the Captain. He rings every coin to find a counterfeit. Not a forger, any way, he mutters; and Jonah is put down for his passage. 'Point out my stateroom, Sir,' says Jonah now. 'I'm travel-weary; I need sleep.' 'Thou look'st like it,' says the Captain, 'there's thy room.'

"Jonah's prodigy of ponderous misery drags him drowning down to sleep.

"And now the time of tide has come; the ship casts off her cables; and from the deserted wharf the uncheered ship for Tarshish, all careening, glides to sea.

That ship, my friends, was the first of recorded smugglers! the contraband was Jonah. But the sea rebels; he will not bear the wicked burden. A dreadful storm comes on, the ship is like to break. In all this raging tumult, Jonah sleeps his hideous sleep. But the frightened master comes to him, and shrieks in his dead ear, 'What meanest thou, O sleeper! arise!' Startled from his lethargy by that direful cry, Jonah staggers to his feet, and stumbling to the deck, grasps a shroud, to look out upon the sea. But at that moment he is sprung upon by a panther billow leaping over the bulwarks.

"Terrors upon terrors run shouting through his soul. In all his cringing attitudes, the God-fugitive is now too plainly known. The sailors mark him; more and more certain grow their suspicions of him, and at last, fully to test the truth, by referring the whole matter to high Heaven, they fall to casting lots, to see for whose cause this great tempest was upon them. The lot is Jonah's; that discovered, then how furiously they mob him with

The Seamen's Bethel, 1870s. *The Bethel opened in 1832 as a place of worship for seamen and others. The building visited by Melville was smaller with no steeple, a staircase on the outside, and the pulpit adjacent to the entrance. To worship, chapel-goers ascended the exterior stairs and entered the chapel facing the congregation. After a fire in 1866, the chapel was rebuilt with the staircase enclosed and the pulpit placed at the opposite end. The first chaplain was the Reverend Enoch Mudge (1832–43), perhaps the inspiration for Father Mapple.*

Kingman Family Collection

their questions. 'What is thine occupation? Whence comest thou? Thy country? What people?'

"'I am a Hebrew,' he cries—and then—'I fear the Lord the God of Heaven who hath made the sea and the dry land!' Fear him, O Jonah? Aye, well mightest thou fear the Lord God *then*! Straightway, he now goes on to make a full confession; whereupon the mariners became more and more appalled, but still are pitiful. When wretched Jonah cries out to them to take him and cast him forth into the sea, for he knew that for his sake this great tempest was upon them; they mercifully turn from him, and seek by other means to save the ship. But all in vain; the indignant gale howls louder; then, with one hand raised invokingly to God, with the other they not unreluctantly lay hold of Jonah.

"And now behold Jonah taken up as an anchor and dropped into the sea; when instantly an oily calmness floats out from the east, and the sea is still, as Jonah carries down the gale with him, leaving smooth water behind. He goes down in the whirling heart of such a masterless commotion that he scarce heeds the moment when he drops seething into the yawning jaws awaiting him; and the whale shoots-to all his ivory teeth, like so many white bolts, upon his prison. Then Jonah prayed unto the Lord out of the fish's belly. But observe his prayer, and so many white bolts, upon his prison. Then Jonah prayed unto learn a weighty lesson. For sinful as he is, Jonah does not weep and wail for direct deliverance. He feels that his dreadful punishment is just. He leaves all his deliverance to God, contenting himself with this, that spite of all his pains and pangs, he will still look towards His holy temple. And here, shipmates, is true and faithful repentance; not clamorous for pardon, but grateful for punishment. Shipmates, I do not place Jonah before you to be copied for his sin but I do place him before you as a model for repentance.

"Jonah, appalled at the hostility he should raise, fled from his mission, and sought to escape his duty and his God by taking ship at Joppa. But God is everywhere; Tarshish he never reached. As we have seen, God came upon him in the whale, and swallowed him down to living gulfs of doom. Then God spake unto the fish; and from the shuddering cold and blackness of the sea, the whale came breeching up towards the warm and pleasant sun, and all the delights of air and earth; and 'vomited out Jonah upon the dry land;' when the word of the Lord came a second time; and Jonah, bruised and beaten—his ears, like two sea-shells, still multitudinously murmuring of the ocean—Jonah did the Almighty's bidding. And what was that, shipmates? To preach the Truth to the face of Falsehood! That was it!

"This, shipmates, this is that other lesson; and woe to that pilot of the living God who slights it. Woe to him who would not be true, even though to be false were salvation! Yea, woe to him who, as the great Pilot Paul has it, while preaching to others is himself a castaway!"

He drooped and fell away from himself for a moment; then lifting his face to them again, showed a deep joy in his eyes, as he cried out with a heavenly enthusiasm,—"But oh! shipmates! Delight is to him, whom all the waves of the billows of the seas of the boisterous mob can never shake from this sure Keel of the Ages. And eternal delight and deliciousness will be his, who coming to lay him down, can say with his final breath—Oh Father—mortal or immortal, here I die. I have striven to be Thine, more than to be this world's, or mine own. Yet this is nothing; I leave eternity to Thee; for what is man that he should live out the lifetime of his God?"

He said no more, but slowly waving a benediction, covered his face with his hands, and so remained kneeling, till all the people had departed, and he was left alone in the place.

Jonah's release, circa 1585. *"Yet even then—out of the belly of hell—when the whale grounded upon the ocean's utmost bones, God heard the engulphed, repenting prophet when he cried."* — Chapter 9. *Engraving by Marten de Vos. Kendall Collection, NBWM.*

Spiritualism at Liberty Hall, 1860. *A brief visit to New Bedford inspired this illustration depicting a woman lecturing on the ills of alcohol to a half-hearted gathering of pious souls at Liberty Hall. At the time, New Bedford was heavily entrenched in Protestant virtue—Quakers, Baptists, Methodists and Congregationalists were well represented. In 1858, Melville returned to New Bedford to give a lecture of his own—a not-so-memorable talk entitled "The Statues of Rome"—before an audience of 400.*
— *Engraving from* Harper's New Monthly Magazine.

Chapter X, XI

A Bosom Friend · Nightgown

Maori carving, 1836.
— Illustration by Alfred T. Agate. From Wilkes: Narrative of the United States.... New Bedford Whaling Museum.

Returning to the Spouter-Inn from the Chapel, I found Queequeg there quite alone; he having left the Chapel before the benediction some time. He was sitting on a bench before the fire, with his feet on the stove hearth, and in one hand was holding close up to his face that little negro idol of his; peering hard into its face, and with a jack-knife gently whittling away at its nose, meanwhile humming to himself in his heathenish way.

But being now interrupted, he put up the image; and pretty soon, going to the table, took up a large book there, and placing it on his lap began counting the pages with deliberate regularity; at every fiftieth page—as I fancied—stopping a moment, looking vacantly around him, and giving utterance to a long-drawn gurgling whistle of astonishment.

With much interest I sat watching him. Savage though he was, and hideously marred about the face—at least to my taste—his countenance yet had a something in it which was by no means disagreeable. You cannot hide the soul. Through all his unearthly tattooings, I thought I saw the traces of a simple honest heart; and in his large, deep eyes, fiery black and bold, there seemed tokens of a spirit that would dare a thousand devils.

Whether it was, too, that his head being shaved, his forehead was drawn out in freer and brighter relief, I would not venture to decide. It may seem ridiculous, but it reminded me of General Washington's head, as seen in the popular busts of him. Queequeg was George Washington cannibalistically developed.

Here was a man some twenty thousand miles from home, by the way of Cape Horn, thrown among people as strange to him as though he were in the planet Jupiter; and yet he seemed entirely at his ease; preserving the utmost serenity; content with his own companionship; always equal to himself.

I drew my bench near him, and made some friendly signs and hints, doing my best to talk with him meanwhile. We then turned over the book together, and I endeavored to explain to him the purpose of the printing, and the meaning of the few pictures that were in it. Thus I soon engaged his interest; and from that we went to jabbering the best we could about the various outer sights to be seen in this famous town. Soon I proposed a social smoke; and, producing his pouch and tomahawk, he quietly offered me a puff. And then we sat exchanging puffs from that wild pipe of his, and keeping it regularly passing between us.

If there yet lurked any ice of indifference towards me in the Pagan's breast, this pleasant, genial smoke we had, soon thawed it out, and left us cronies. When our smoke was over, he pressed his forehead against mine, clasped me around the waist, and said that henceforth we were married; meaning that we were bosom friends; he would gladly die for me, if need should be.

After supper, and another social chat and smoke, we went to our room together. He then went about his evening prayers, took out his idol, and removed the paper fireboard. By certain signs and symptoms, I thought he seemed anxious for me to join him; but well knowing what was to follow, I deliberated a moment whether, in case he invited me, I would comply or otherwise.

I was a good Christian; born and bred in the bosom of the infallible Presbyterian Church. How then could I unite with this wild idolator in worshipping his piece of wood? Now, Queequeg is my fellow man. And what do I wish that this Queequeg would do to me? Why, unite with me in my particular Presbyterian form of worship. consequently, I must then unite with him in his; ergo, I must turn idolator. So I kindled the shavings; helped prop up the innocent little idol; offered him burnt biscuit with Queequeg; salamed before him twice or thrice; kissed his nose; and that done, we undressed and went to bed, at peace with our own consciences and all the world. But we did not go to sleep without some little chat.

How it is I know not; but there is no place like a bed for confidential disclosures between friends. Man and wife, they say, there open the very bottom of their souls to each other; and some old couples often lie and chat over old times till nearly morning. Thus, then, in our hearts' honeymoon, lay I and Queequeg—a cosy, loving pair.

Maori Chief Kotowatowa of Whangara, New Zealand, 1836. The Maori were encountered by Lt. Charles Wilkes of the U.S. Exploration Expedition, who is given credit for the discovery of various lands in the South Pacific, particularly in Antarctica. According to Australian historian Peter Macinnis, the eccentric Wilkes "seems to have been, at least in part, the model on whom Herman Melville based his character, Captain Ahab." —
Illustration by Alfred T. Agate. From Wilkes: Narrative of the United States.... New Bedford Whaling Museum.

Chapter XII

Biographical

A War Canoe of New Zealand. — *Kendall Collection, NBWM.*

Queequeg was a native of Kokovoko, an island far away to the West and South. It is not down in any map; true places never are.

When a new-hatched savage running wild about his native woodlands in a grass clout, followed by the nibbling goats, as if he were a green sapling; even then, in Queequeg's ambitious soul, lurked a strong desire to see something more of Christendom than a specimen whaler or two. His father was a High Chief, a King; his uncle a High Priest. There was excellent blood in his veins—royal stuff.

A Sag Harbor ship visited his father's bay, and Queequeg sought a passage to Christian lands. But the ship, having her full complement of seamen, spurned his suit; and not all the King his father's influence could prevail. But Queequeg vowed a vow. Alone in his canoe, he paddled off to a distant strait, which he knew the ship must pass through when she quitted the island. On one side was a coral reef; on the other a low tongue of land, covered with mangrove thickets that grew out into the water. Hiding his canoe, still afloat, among these thickets, with its prow seaward, he sat down in the stern, paddle low in hand; and when the ship was gliding by, like a flash he darted out; gained her side; with one backward dash of his foot capsized and sank his canoe; climbed up the chains; and throwing himself at full length upon the deck, grappled a ring-bolt there, and swore not to let it go, though hacked in pieces.

In vain the captain threatened to throw him overboard; suspended a cutlass over his naked wrists; Queequeg was the son of a King, and Queequeg budged not. The captain at last relented, and told him he might make himself at home. But this fine young savage—this sea Prince of Wales, never saw the captain's cabin. They put him down among the sailors, and made a whaleman of him. But, alas! the practices of whalemen soon convinced him that even Christians could be both miserable and wicked; infinitely more so, than all his father's heathens. Arrived at last in old Sag Harbor; and seeing what the sailors did there; and then going on to Nantucket, and seeing how they spent their wages in *that* place also, poor Queequeg gave it up for lost. Thought he, it's a wicked world in all meridians; I'll die a pagan.

And thus an old idolator at heart, he yet lived among these Christians, wore their clothes, and tried to talk their gibberish. Hence the queer ways about him, though now some time from home.

But by and by, he said, he would return,—as soon as he felt himself baptized again. For the nonce, however, he proposed to sail about, and sow his wild oats in all four oceans. They had made a harpooneer of him, and that barbed iron was in lieu of a sceptre now.

I asked him what might be his immediate purpose, touching his future movements. He answered, to go to sea again, in his old vocation. Upon this, I told him that whaling was my own design, and informed him of my intention to sail out of Nantucket, as being the most promising port for an adventurous whaleman to embark from. He at once resolved to accompany me to that island, ship aboard the same vessel, get into the same watch, the same boat, the same mess with me, in short to share my every hap; with both my hands in his, boldly dip into the Potluck of both worlds. To all this I joyously assented; for besides the affection I now felt for Queequeg, he was an experienced harpooneer, and as such, could not fail to be of great usefulness to one, who, like me, was wholly ignorant of the mysteries of whaling, though well acquainted with the sea, as known to merchant seamen.

His story being ended with his pipe's last dying puff, Queequeg embraced me, pressed his forehead against mine, and blowing out the light, we rolled over from each other, this way and that, and very soon were sleeping.

New Bedford Whaling Museum

Whangara, New Zealand, 1836. *This is the island village of the Maori high chief, Kotowatowa, 30 miles south of Cape Brett. By the end of the 19th century, Maori numbers had dwindled to 40,000. However, through the efforts of their own chiefs, they have reemerged as an economically self-sufficient minority in New Zealand, numbering more than 500,000.* — Illustration by Alfred T. Agate. From Wilkes: Narrative of the United States….

Chapter XIII

Wheelbarrow

Taber's Wharf, New Bedford, 1860. — *Kingman Family Collection*

Next morning, Monday, after disposing of the embalmed head to a barber, for a block, I settled my own and comrade's bill; using, however, my comrade's money. The grinning landlord, as well as the boarders, seemed amazingly tickled at the sudden friendship which had sprung up between me and Queequeg—especially as Peter Coffin's cock and bull stories about him had previously so much alarmed me concerning the very person whom I now companied with.

We borrowed a wheelbarrow, and embarking our things, including my own poor carpet-bag, and Queequeg's canvas sack and hammock, away we went down to "the Moss," the little Nantucket packet schooner moored at the wharf. As we were going along the people stared; not at Queequeg so much—for they were used to seeing cannibals like him in their streets,—but at seeing him and me upon such confidential terms. But we heeded them not, going along wheeling the barrow by turns, and Queequeg now and then stopping to adjust the sheath on his harpoon barbs. I asked him why he carried such a troublesome thing with him ashore, and whether all whaling ships did not find their own harpoons. To this, in substance, he replied, that though what I hinted was true enough, yet he had a particular affection for his own harpoon, because it was of assured stuff, well tried in many a mortal combat, and deeply intimate with the hearts of whales. In short, like many inland reapers and mowers, who go into the farmers' meadows armed with their own scythes—though in no wise obliged to furnished them—even so, Queequeg, for his own private reasons, preferred his own harpoon.

Shifting the barrow from my hand to his, he told me a funny story about the first wheelbarrow he had ever seen. It was in Sag Harbor. The owners of his ship, it seems, had lent him one, in which to carry his heavy

Merchants Wharf, 1882. *The bark* Canton *(center), unloads part of her hefty cargo of 1,950 barrels of sperm oil after a 4-year voyage to the Pacific.*

Spinner Collection

chest to his boarding house. Not to seem ignorant about the thing—though in truth he was entirely so, concerning the precise way in which to manage the barrow—Queequeg puts his chest upon it; lashes it fast; and then shoulders the barrow and marches up the wharf. "Why," said I, "Queequeg, you might have known better than that, one would think. Didn't the people laugh?"

Upon this, he told me another story. The people of his island of Rokovoko, it seems, at their wedding feasts express the fragrant water of young cocoanuts into a large stained calabash like a punchbowl; and this punchbowl always forms the great central ornament on the braided mat where the feast is held. Now a certain grand merchant ship once touched at Rokovoko, and its commander—from all accounts, a very stately punctilious gentleman, at least for a sea captain—this commander was invited to the wedding feast of Queequeg's sister, a pretty young princess just turned of ten. Well; when all the wedding guests were assembled at the bride's bamboo cottage, this Captain marches in, and being assigned the post of honor, placed himself over against the punchbowl, and between the High Priest and his majesty the King, Queequeg's father. Grace being said,—for those people have their grace as well as we—though Queequeg told me that unlike us, who at such times look downwards to our platters, they, on the contrary, copying the ducks, glance upwards to the great Giver of all feasts—Grace, I say, being said, the High Priest opens the banquet by the immemorial ceremony of the island; that is, dipping his consecrated and consecrating fingers into the bowl before the blessed beverage circulates. Seeing himself placed next the Priest, and noting the ceremony, and thinking himself—being Captain of a ship—as having plain precedence over a mere island King, especially in the King's own house—the Captain coolly proceeds to wash his hands in the punch bowl;—taking it I suppose for a huge finger-glass. "Now," said Queequeg, "what you tink now,—Didn't our people laugh?"

William R. Hegarty Collection

William R. Hegarty Collection

Spinner Collection

Ship caulkers and carpenters, 1915. *The whaling bark* Greyhound *gets her seams caulked before her next voyage.*

Coopers on New Bedford wharf, circa 1910.

Ship carpenters, 1900. *The ornate bowsprit of the* Morning Star *overlooks two elderly carpenters working on a ship's crossbeams.*

Many dockworkers had once been whalers themselves. "They were nearly all whalemen; …sea carpenters, and sea coopers, and sea blacksmiths, and harpooneers, and ship keepers; a brown and brawny company, with bosky beards; an unshorn, shaggy set, all wearing monkey jackets for morning gowns." — Chapter 5. *Toward the end of the whaling era, dockworkers such as carpenters, coopers and shipkeepers were mostly elderly men. The industry and the old wooden ships were relics of the past century—not part of a young man's future.*

Bark Tamerlane *outbound through icy harbor, 1888.* *When Melville's ship, Acushnet, finally pushed off from Fairhaven's Union Wharf on January 3, 1841, she may have faced a similar scene—an inner harbor clogged with ice. Bark Tamerlane, bound for the Arctic in this photograph, had been known as a lucky ship, bringing in handsome profits for her owners. Her luck changed, however, when she ran into lava cliffs on the Hawaiian coast near Hilo in 1892 and sank. Captain W. F. Howland, who created a scandal when he and his First Mate abandoned ship first, perished along with 18 men.* — Joseph G. Tirrell photograph. New Bedford Whaling Museum.

New Bedford Harbor, 1835. *View from Fairhaven shows an array of winter activity in a landscape blanketed in snow. After signing his shipping papers on December 25, Melville had a week to himself in port while the ship fitted out—time to visit the Bethel and experience the street. As for the weather, Melville may have witnessed the same driving winds, snow and sleet that ripped through the Labrador whalemen, the Bethel parishioners and Ishmael himself. According to the log of the ship* Charles, *in port at the time, conditions on December 27 are described: "First part commences with a thick snow storm and wind @ S. East." Searching for lodging, Ishmael trudges by the Swordfish Inn where light from the window came in "such fervent rays, that it seemed to have melted the packed snow and ice from before the house, for everywhere else the congealed frost lay ten inches thick in a hard, asphaltic pavement."* — Chapter 2. Anonymous watercolor. Weather information from Dahl: Melville Society Extracts.

Kendall Collection, NBWM

"Coming to Anchor," 1873. Whaleships in New Bedford Harbor with Fairhaven in background. — *Watercolor by Benjamin Russell.*

"Whaleship Sharon of Fairhaven," 1845. In April 1841, the Sharon hoisted anchor and embarked on a fateful voyage in which she fell victim to a violent mutiny in the South Pacific (see page 190). — *Painting attributed to Caleb Purrington.*

Forbes Collection of the MIT Museum

Forbes Collection of the MIT Museum

At last, passage paid, and luggage safe, we stood on board the schooner. Hoisting sail, it glided down the Acushnet river. On one side, New Bedford rose in terraces of streets, their ice-covered trees all glittering in the clear, cold air. Huge hills and mountains of casks on casks were piled upon her wharves, and side by side the world-wandering whale ships lay silent and safely moored at last; while from others came a sound of carpenters and coopers, with blended noises of fires and forges to melt the pitch, all betokening that new cruises were on the start; that one most perilous and long voyage ended, only begins a second; and a second ended, only begins a third, and so on, for ever and for aye. Such is the endlessness, yea, the intolerableness of all earthly effort.

Gaining the more open water, the bracing breeze waxed fresh; the little Moss tossed the quick foam from her bows, as a young colt his snortings. How I snuffed that Tartar air!—how I spurned that turnpike earth!—that common highway all over dented with the marks of slavish heels and hoofs; and turned me to admire the magnanimity of the sea which will permit no records.

At the same foam-fountain, Queequeg seemed to drink and reel with me. His dusky nostrils swelled apart; he showed his filed and pointed teeth. On, on we flew,

The Panorama. The panorama was the cinema of the 19th century. Invented independently by several European painters in the 1780s, panoramas are considered the first visual "mass media," taking viewers to exotic lands and historical events. Early panoramas, or cycloramas, displayed in specially built rotundas, gave spectators the feeling they were seeing the real thing. The artist John Vanderlyn introduced the 360-degree panorama to America in 1818 but discovered audiences found them static and too limiting. More appealing was the moving panorama—a long narrative painting rolled across a stage between two upright reels to simulate a lengthy journey through the American landscape. The moving panorama was more portable and could be exhibited easily in different cities. These moving panoramas introduced thousands of Americans to the majesty of their new nation and exotic lands around the world. — Text from New Bedford Whaling Museum exhibit.

Scenes from "Whaling Voyage Round the World," 1845. In 1841, at the age of 37, Benjamin Russell left his wife and three children to spend the next three-and-a-half years aboard the whaleship Kutusoff on a voyage around the world. He hoped to earn enough to clear debts suffered in the banking crisis of 1832-33. He earned $894.51, barely enough to pay his debts, but the trip inspired him to begin a career as a marine artist. He transformed the sketches he made at sea into lovely watercolors and oil paintings. Later he teamed up with local sign painter Caleb Purrington and together they produced the 1,500-foot panorama entitled "Whaling Voyage Round the World," an extravaganza, even in its day. — Text from New Bedford Whaling Museum exhibit.

In the winter of 1849, Melville visited Boston at the same time that Russell and Purrington's panorama was enjoying a well-publicized showing. At this time, Melville was just beginning work on Moby-Dick and it's quite possible he saw the showing.

"The hall has been crowded every evening," wrote one reader in a letter to the New Bedford Mercury (December 6, 1848). "All who were present, and there were not a few who were well qualified by actual observation to judge of its merits, allow that it came fully up to the mark, in its delineation of scenes well known to our gallant whalers…"

The Voyage begins. The canvas moves from right to left, therefore the voyage begins at far right (top) at the New Bedford end of the Fairhaven bridge. As we pan to the left across the inner harbor, we see recreational boats, small fishing boats and a Revenue cutter. The bustling harbor activity is set before an accurate rendition of New Bedford's skyline—a scene reminiscent for Melville, and one that would certainly have inspired him had he been in attendance. As the panorama continues (bottom, right to left), we see a whaleship in stream, a merchant brig, inshore fishing boats, Palmer's Island, a large salt works complex, farm houses, groves of trees, coastal schooners and packets.

New Bedford Whaling Museum

Leaving the harbor, 1845. *At the southern tip of New Bedford's peninsula, extending into Buzzards Bay, is the Clark's Point lighthouse (center). The prominent building at far right, surrounded by tree groves, is the city's alms house or "Poor Farm." Under sail is an inbound coaster, and next, the whaleship* Janus. *In the foreground, the whaleship* Niger *returns from a voyage under tow of the steamer* Massachusetts; *and a pilot boat brings up the rear.* — Russell & Purrington panorama. New Bedford Whaling Museum.

and our offing gained, the Moss did homage to the blast; ducked and dived her brows as a slave before the Sultan. Sideways leaning, we sideways darted; every ropeyarn tingling like a wire; the two tall masts buckling like Indian canes in land tornadoes. So full of this reeling scene were we, as we stood by the plunging bowsprit, that for some time we did not notice the jeering glances of the passengers, a lubber-like assembly, who marvelled that two fellow beings should be so companionable; as though a white man were anything more dignified than a whitewashed negro. But there were some boobies and bumpkins there, who, by their intense greenness, must have come from the heart and centre of all verdure. Queequeg caught one of these young saplings mimicking him behind his back. I thought the bumpkin's hour of doom was come. Dropping his harpoon, the brawny savage caught him in his arms, and by an almost miraculous dexterity and strength, sent him high up bodily into the air; then slightly tapping his stern in mid-somerset, the fellow landed with bursting lungs upon his feet, while Queequeg, turning his back upon him, lighted his tomahawk pipe and passed it to me for a puff.

"Hallo, *you* sir," cried the Captain, a gaunt rib of the sea, stalking up to Queequeg, "what in thunder do you mean by that? Don't you know you might have killed that chap?"

"Kill-e," cried Queequeg, twisting his tattooed face into an unearthly expression of disdain, "ah! him bevy small-e fish-e; Queequeg no kill-e so small-e fish-e; Queequeg kill-e big whale!"

"Look you," roared the Captain, "I'll kill-e *you*, you cannibal, if you try any more of your tricks aboard here; so mind your eye."

But it so happened just then, that it was high time for the Captain to mind his own eye. The prodigious strain upon the main-sail had parted the weather-sheet, and the tremendous boom was now flying from side to side, completely sweeping the entire after part of the deck. The poor fellow whom Queequeg had handled so roughly, was swept overboard; all hands were in a panic; and to attempt snatching at the boom to stay it, seemed madness. It flew from right to left, and back again, almost in one ticking of a watch, and every instant seemed on the point of snapping into splinters. Nothing

To the sea, 1845. *As the panorama takes us to outer bay and seaward, we see the whaleship* India *(right) surrounded by sail boats and fishing boats. At left, a Chinese junk labors inbound, and the tugboat* R. B. Forbes *heads outbound. In the background is Mishaum Point, Dartmouth.*

New Bedford Whaling Museum

was done, and nothing seemed capable of being done; those on deck rushed towards the bows, and stood eyeing the boom as if it were the lower jaw of an exasperated whale. In the midst of this consternation, Queequeg dropped deftly to his knees, and crawling under the path of the boom, whipped hold of a rope, secured one end to the bulwarks, and then flinging the other like a lasso, caught it round the boom as it swept over his head, and at the next jerk, the spar was that way trapped, and all was safe. The schooner was run into the wind, and while the hands were clearing away the stern boat, Queequeg, stripped to the waist, darted from the side with a long living arc of a leap. For three minutes or more he was seen swimming like a dog, throwing his long arms straight out before him, and by turns revealing his brawny shoulders through the freezing foam. I looked at the grand and glorious fellow, but saw no one to be saved. The greenhorn had gone down. Shooting himself perpendicularly from the water, Queequeg now took an instant's glance around him, and seeming to see just how matters were, dived down and disappeared. A few minutes more, and he rose again, one arm still striking out, and with the other dragging a lifeless form. The boat soon picked them up. The poor bumpkin was restored. All hands voted Queequeg a noble trump. Was there ever such unconsciousness? He did not seem to think that he at all deserved a medal from the Humane and Magnanimous Societies. He only asked for water—fresh water—something to wipe the brine off; that done, he put on dry clothes, lighted his pipe, and leaning against the bulwarks, and mildly eyeing those around him, seemed to be saying to himself—"It's a mutual, joint-stock world, in all meridians. We cannibals must help these Christians."

Whaleship Gratitude *returning from a voyage, 1845.* *Back from the Indian Ocean, the* Gratitude *sails into New Bedford harbor laden with 2,218 barrels of oil and 9,000 pounds of whalebone. This watercolor is one of the earliest known ship portraits by Benjamin Russell, painted around the time Russell and Purrington were creating their panorama.*

Forbes Collection of the MIT Museum

Herman Melville in the Pacific, 1842-1844. Melville deserted the Acushnet in 1842 after 18 months at sea, unable to continue with the irascible, Ahab-like Captain Pease. He and shipmate Tobias Greene deserted at Nuka Hiva Island in the Marquesas by slipping over the side of the ship during a downpour. The two men headed for the jungle, climbed volcanic mountains and steep gorges before reaching the fertile valley of the Typees, a people rumored to be cannibals. Melville, suffering from fever, chills and an infected leg, persuaded the Polynesians to send Toby for medicine. Alas, Toby was shanghaied (or voluntarily boarded) an Australian whaler and never returned.

Melville stayed in a bamboo hut to recuperate and the natives treated him kindly, providing him with coconuts, breadfruit and the company of young Polynesian women. Though he admired the simple life and open sensuality of the society, he felt like a prisoner. After four weeks he escaped to Taiohae Bay on the southern coast of the island.

Melville's first book, "Typee: A Peep at Polynesian Life" was a fictional travelogue based on his adventures on the islands. It became an international best seller, in part, because of the sexual escapades of the hero. Thus Melville's career was launched.

At Taiohae Bay, Melville signed on to the Lucy Ann, an old whaling ship from Sydney with a sickly captain, a drunken first mate and a mutinous crew riddled with venereal disease. He did, however, become friends with the ship's steward, John Troy, nicknamed "Dr. Long Ghost." When the Lucy Ann arrived in Tahiti, Melville and 11 mates refused to continue on the unsafe ship.

Sailors from a nearby French frigate boarded the ship and arrested the mutineers, including Melville, who was confined for two days in "double irons" and threatened with a trial that proved to be a bluff. Melville, Troy and the others were locked up in the "Calabooza Beretanee," a makeshift outdoor jail. Soon he and Troy made friends with the warden who let them run loose all day if they promised to return to the jail at night.

While in the South Pacific islands, Melville saw first-hand the destructive influence of colonialism and Christian missionaries, and developed critical attitudes that found their way into his writing. Troy and Melville escaped by canoe to the island of Imeeo and worked as farmhands on a plantation. By 1843, Melville was eager to return home.

Melville's true bosom friend aboard *Acushnet* was Toby Greene, who deserted ship with him in 1842. — *Daguerreotype, 1860. New Bedford Whaling Museum.*

Melville signed up on another whaler, the Charles & Henry of Nantucket, hoping to wend his way home. After three months, however, they caught no whales. When the ship put in at Lahaina, on the island of Maui, Melville quit the voyage. He had arrived during a time of unrest, and after a period working as a bookkeeper, it was time to go. He enlisted as an ordinary seaman in the United States Navy so he could return home in the frigate United States, bound for Boston.

The atmosphere was oppressive, the crew was overworked and underfed. However, the crew was large enough to include some aspiring writers and poets. An Englishman, Jack Chase, had great influence on Melville, nurturing him and teaching him seamanship. Having been away four years, Melville returned home in the fall of 1844 and announced he had completed "my Harvard and my Yale."

Tiring of life in paradise, Melville ran away from the Typees and was picked up by the Lucy Ann—an ill-fated ship with a sickly captain and mutinous crew. — From Chase journal, whaleship Massachusetts, 1850. Kendall Collection, NBWM.

A warrant for the arrest of seven deserters of the ship Acushnet, issued in 1843, includes Herman Melville and Tobias Greene. The outlaws jumped ship at Nuka Hiva just 18 months into their voyage. — New Bedford Whaling Museum.

Nuka Hiva, Marquesas Islands, 1845. This view from Russell and Purrington's 1,500-foot panorama shows French men-of-war and whaleships at anchor in Typee Bay. The French took possession of the Marquesas in 1842, shortly after Melville and Greene jumped ship there in July. Visible on land is a fort, the King's residence, the Governor's house, a hospital and barracks. It's possible that at the time Russell visited these waters aboard the Kutusoff and sketched it, Melville was in the backlands eating breadfruit from the hands of a Marquesan maiden. — New Bedford Whaling Museum.

Chapter XIV

Nantucket

Nothing more happened on the passage worthy the mentioning; so, after a fine run, we safely arrived in Nantucket.

View of Nantucket Harbor from the church tower, 1872. —Scribner's Magazine.

Nantucket! Take out your map and look at it. See what a real corner of the world it occupies; how it stands there, away off shore. Look at it—a mere hillock, and elbow of sand; all beach, without a background. There is more sand there than you would use in twenty years as a substitute for blotting paper. Some gamesome wights will tell you that they have to plant weeds there, they don't grow naturally; that they import Canada thistles; that they have to send beyond seas for a spile to stop a leak in an oil cask; that pieces of wood in Nantucket are carried about like bits of the true cross in Rome; that people there plant toad-stools before their houses, to get under the shade in summer time; that one blade of grass makes an oasis, three blades in a day's walk a prairie; that they wear quicksand shoes, something like Laplander snowshoes; that they are so shut up, belted about, every way inclosed, surrounded, and made an utter island of by the ocean, that to their very chairs and tables small clams will sometimes be found adhering, as to the backs of sea turtles.

Let America add Mexico to Texas, and pile Cuba upon Canada; let the English overswarm all India, and hang out their blazing banner from the sun; two thirds of this terraqueous globe are the Nantucketer's. For the sea is his; he owns it, as Emperors own empires. Merchant ships are but extension bridges; armed ones but floating forts; even pirates and privateers, though following the sea as highwaymen the road, they but plunder other ships, other fragments of the land like themselves, without seeking to draw their living from the bottomless deep itself. The Nantucketer, he alone resides and riots on the sea; he alone, in Bible language, goes down to it in ships; to and fro ploughing it as his own special plantation. He lives on the sea, as prairie cocks in the prairie; he hides among the waves, he climbs them as chamois hunters climb the Alps. For years he knows not the land; so that when he comes to it at last, it smells like another world, more strangely than the moon would to an Earthsman. With the landless gull, that at sunset folds her wings and is rocked to sleep between billows; so at nightfall, the Nantucketer, out of sight of land, furls his sails, and lays him to his rest, while under his very pillow rush herds of walruses and whales.

Town of Sherbourne, Nantucket, 1815. "What wonder, then, that these Nantucketers, born on a beach, should take to the sea for a livelihood! They first caught crabs and quahogs in the sand; grown bolder, they waded out with nets for mackerel; more experienced, they pushed off in boats and captured cod; and at last, launching a navy of great ships on the sea, explored this watery world; …and in all seasons and all oceans declared everlasting war with the mightiest animated mass that has survived the flood."

Kendall Collection, NBWM

Chapter XV

Chowder

Lighthouse at Sankaty Head, 1876. — *From Davis:* Nimrod of the Sea.

It was quite late in the evening when the little Moss came snugly to anchor, and Queequeg and I went ashore; so we could attend to no business that day, at least none but a supper and a bed. The landlord of the Spouter-Inn had recommended us to his cousin Hosea Hussey of the Try Pots, whom he asserted to be the proprietor of one of the best kept hotels in all Nantucket, and moreover he had assured us that cousin Hosea, as he called him, was famous for his chowders. In short, he plainly hinted that we could not possibly do better than try pot-luck at the Try Pots. By dint of beating about a little in the dark, and now and then knocking up a peaceable inhabitant to inquire the way, we at last came to something which there was no mistaking.

Two enormous wooden pots painted black, and suspended by asses' ears, swung from the cross-trees of an old top-mast, planted in front of an old doorway. The horns of the cross-trees were sawed off on the other side, so that this old top-mast looked not a little like a gallows. Perhaps I was over sensitive to such impressions at the time, but I could not help staring at this gallows with a vague misgiving.

I was called from these reflections by the sight of a freckled woman with yellow hair and a yellow gown, standing in the porch of the inn, under a dull red lamp swinging there, that looked much like an injured eye, and carrying on a brisk scolding with a man in a purple woollen shirt.

Upon making known our desires for a supper and a bed, Mrs. Hussey, postponing further scolding for the present, ushered us into a little room, and seating us at a table spread with the relics of a recently concluded repast, turned round to us and said—"Clam or Cod?"

"A clam for supper? a cold clam; is *that* what you mean, Mrs. Hussey?" says I; "but that's a rather cold and clammy reception in the winter time, ain't it, Mrs. Hussey?"

But being in a great hurry to resume scolding the man in the purple shirt, who was waiting for it in the entry, and seeming to hear nothing but the word "clam," Mrs. Hussey hurried towards an open door leading to the kitchen, and bawling out "clam for two," disappeared.

"Queequeg," said I, "do you think that we can make out a supper for us both on one clam?"

Arrival of ship Three Brothers, Nantucket, 1841. *The 384-ton ship* Three Brothers, *captained by Henry Phelone, returned from the Pacific in April 1841 carrying 2,719 barrels of sperm oil. The American whaling fleet in 1840 listed 675 vessels manned with 16,000–17,000 men. Over 400 of these ships hailed from southeastern New England.* — *Color lithograph based on a painting by Oswald Brett.*

Kendall Collection, NBWM

However, a warm savory steam from the kitchen served to belie the apparently cheerless prospect before us. But when that smoking chowder came in, the mystery was delightfully explained. It was made of small juicy clams, scarcely bigger than hazel nuts, mixed with pounded ship biscuit, and salted pork cut up into little flakes; the whole enriched with butter, and plentifully seasoned with pepper and salt. Our appetites being sharpened by the frosty voyage, and in particular, Queequeg seeing his favorite fishing food before him, and the chowder being surpassingly excellent, we despatched it with great expedition: when leaning back a moment and bethinking me of Mrs. Hussey's clam and cod announcement, I thought I would try a little experiment. Stepping to the kitchen door, I uttered the word "cod" with great emphasis, and resumed my seat. In a few moments the savory steam came forth again, but with a different flavor, and in good time a fine cod-chowder was placed before us.

We resumed business; and while plying our spoons in the bowl, thinks I to myself, I wonder now if this here has any effect on the head? What's that stultifying saying about chowder-headed people? "But look, Queequeg, ain't that a live eel in your bowl? Where's your harpoon?"

Fishiest of all fishy places was the Try Pots, which well deserved its name; for the pots there were always boiling chowders. Chowder for breakfast, and chowder for dinner, and chowder for supper, till you began to look for fish-bones coming through your clothes. The area before the house was paved with clam-shells. Mrs. Hussey wore a polished necklace of codfish vertebrae; and Hosea Hussey had his account books bound in superior old shark-skin. There was a fishy flavor to the milk, too, which I could not at all account for, till one morning happening to take a stroll along the beach among some fishermen's boats, I saw Hosea's brindled cow feeding on fish remnants, and marching along the sand with each foot in a cod's decapitated head, looking very slip-shod, I assure ye.

Supper concluded, we received a lamp, and directions from Mrs. Hussey concerning the nearest way to bed; but, as Queequeg was about to precede me up the stairs, the lady reached forth her arm, and demanded his harpoon; she allowed no harpoon in her chambers. "Why not?" said I; "every true whaleman sleeps with his harpoon—but why not?" "Because it's dangerous," says she. "Ever since young Stiggs coming from that unfort'nt v'y'ge of his, when he was gone four years and a half, with only three barrels of *ile*, was found dead in my first floor back, with his harpoon in his side; ever since then I allow no boarders to take sich dangerous weepons in their rooms at night. So, Mr. Queequeg" (for she had learned his name), "I will just take this here iron, and keep it for you till morning. But the chowder; clam or cod to-morrow for breakfast, men?"

"Both," says I; "and let's have a couple of smoked herring by way of variety."

Macy House, Nantucket, built in 1654. *"The whale-killing historian and poet, Obed Macy, says very truly, that 'The sea to mariners is but a highway: to the whaler it is his field of harvest; it is the home of his business.' "* — Davis: Nimrod of the Sea.

Spinner Collection

Chapter XVI

The Ship

Drawing of the *Acushnet* by Henry Johnson, boat-steerer, on the ship's second voyage, cira 1846.

In bed we concocted our plans for the morrow. But to my surprise and no small concern, Queequeg now gave me to understand, that he had been diligently consulting Yojo—the name of his black little god—and Yojo had told him two or three times over, and strongly insisted upon it everyway, that instead of our going together among the whaling-fleet in harbor, and in concert selecting our craft; instead of this, I say, Yojo earnestly enjoined that the selection of the ship should rest wholly with me, inasmuch as Yojo purposed befriending us; and, in order to do so, had already pitched upon a

Central Wharf, 1914. Andrew Hicks *and* Morning Star *dry their sails after returning from whaling in the Atlantic.* — *Joseph S. Martin photograph.*

William R. Hegarty Collection

vessel, which, if left to myself, I, Ishmael, should infallibly light upon, for all the world as though it had turned out by chance; and in that vessel I must immediately ship myself, for the present irrespective of Queequeg.

Next morning early, leaving Queequeg shut up with Yojo in our little bedroom, I sallied out among the shipping. After much prolonged sauntering and many random inquiries, I learnt that there were three ships up for three-years' voyages—The Devil-Dam, the Tit-bit, and the Pequod. I peered and pryed about the Devil-Dam; from her, hopped over to the Tit-bit; and, finally, going on board the Pequod, looked around her for a moment, and then decided that this was the very ship for us.

You may have seen many a quaint craft in your day, for aught I know;—squared-toed luggers; mountainous Japanese junks; butter-box galliots, and what not; but take my word for it, you never saw such a rare old craft as this same rare old Pequod. She was a ship of the old school, rather small if anything; with an old fashioned claw-footed look about her. Long seasoned and weather-stained in the typhoons and calms of all four oceans, her old hull's complexion was darkened like a French grenadier's, who has alike fought in Egypt and Siberia. Her venerable bows looked bearded. Her masts—cut somewhere on the coast of Japan, where her original ones were lost overboard in a gale—her masts stood stiffly up like the spines of the three old kings of Cologne. Her ancient decks were worn and wrinkled, like the pilgrim-

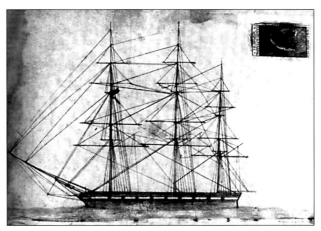

***Whaleship* Acushnet, 1845.** *Desertions from the* Acushnet *began even before she set sail, as two crew members didn't show up. Once under weigh, however, things got better. En route to the Pacific and just 10 weeks under sail, she pulled into Rio de Janeiro and transferred 150 barrels of sperm oil. In April, she rounded Cape Horn and headed to the whaling grounds around the Galapagos Islands. By the time Melville jumped ship after 18 months of whaling,* Acushnet *had only 700 barrels of oil, and by June 1843, she registered 7 desertions. By May 1845, after four and a half years at sea, she returned to Fairhaven with just 850 barrels of sperm oil, 1,350 barrels of whale oil, and 13,500 pounds of whalebone.*

Acushnet *began her second voyage in 1845, a three-year sojourn that brought 1,300 barrels of oil and 10,000 pounds of whalebone. The trip was cut short after five of the crew were lost when a boat was stove by a whale. Never a very lucky ship, the Acushnet's luck bottomed out during her third voyage. On August 16, 1851, she ran aground and was abandoned in the ice off Saint Lawrence Island in the Bering Sea.* — *Watercolor drawing from the journal of boatsteerer Henry Johnson, who perished at sea in 1847.*

***Model/painting of* Acushnet, 1933.** *The* Acushnet *was built under the supervision of master carpenter Ansel Weeks at the Barstow yard in the Mattapoisett section of Rochester in 1840. A fairly large vessel at about 358 tons, 104 ft. long, 28 ft. wide, and 14 ft. deep, the* Acushnet *had two decks, three masts, a square stern, a billethead, but no galleries. She was registered in Fairhaven and captained by Valentine Pease, Jr. of Edgartown. Although no pictures of her exist, the* Acushnet *is thought to have been nearly identical to the* Charles W. Morgan *(see facing page).*
— *Watercolor with string rigging by Richard Noble, Federal Arts Project.*

New Bedford Free Public Library

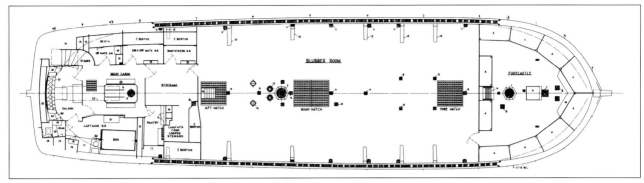

Deck plan of the Charles W. Morgan. — *National Archives.*

worshipped flag-stone in Canterbury Cathedral where Beckett bled. But to all these her old antiquities, were added new and marvellous features, pertaining to the wild business that for more than half a century she had followed. Old Captain Peleg, many years her chief-mate, before he commanded another vessel of his own, and now a retired seaman, and one of the principal owners of the Pequod,—this old Peleg, during the term of his chief-mateship, had built upon her original grotesqueness, and inlaid it, all over, with a quaintness both of material and device, unmatched by anything except it be Thorkill-Hake's carved buckler or bedstead. She was apparelled like any barbaric Ethiopian emperor, his neck heavy with pendants of polished ivory. She was a thing of trophies. A cannibal of a craft, tricking herself forth in the chased bones of her enemies. All round, her unpanelled, open bulwarks were garnished like one continuous jaw, with the long sharp teeth of the sperm whale, inserted there for pins, to fasten her old hempen thews and tendons to. Those thews ran not through base blocks of land wood, but deftly travelled over sheaves of sea-ivory. Scorning a turnstile wheel at her reverend helm, she sported there a tiller; and that tiller was in one mass, curiously carved from the long narrow lower jaw of her hereditary foe.

Broadside view of the Charles W. Morgan. *Built at the Hillman shipyard in New Bedford during the same frigid January when Melville sailed, the* Charles W. Morgan's *specifications are nearly identical to those of the* Acushnet. *Now berthed at Mystic Seaport, Connecticut, the* Morgan *is the last surviving wooden whaleship.*

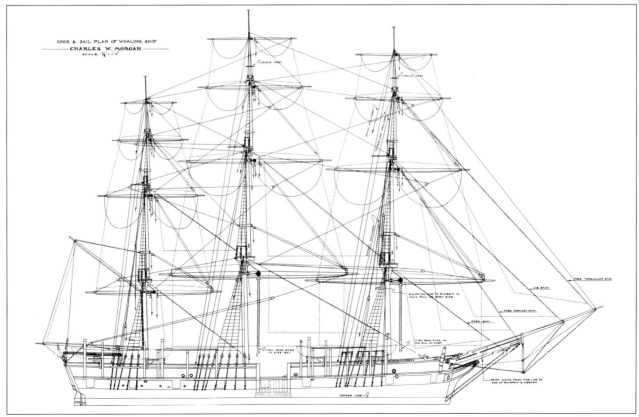

National Archives

Deck of the whaling brig Sullivan, *1900. The crew stands by while the ship undergoes inspection by her agent.* — Kendall Collection, NBWM.

Now when I looked about the quarter-deck, for some one having authority, in order to propose myself as a candidate for the voyage, I could not well overlook a strange sort of tent, or rather wigwam, pitched a little behind the main-mast. A triangular opening faced towards the bows of the ship, so that the insider commanded a complete view forward.

And half concealed in this queer tenement, I at length found one who by his aspect seemed to have authority; and who was now enjoying respite from the burden of command. He was seated on an old-fashioned oaken chair, wriggling all over with curious carving; and the bottom of which was formed of a stout interlacing of the same elastic stuff of which the wigwam was constructed.

"Is this the Captain of the Pequod?" said I, advancing to the door of the tent.

"Supposing it be the Captain of the Pequod, what dost thou want of him?" he demanded.

"I was thinking of shipping."

"Thou wast, wast thou? I see thou are no Nantucketer—ever been in a stove boat?"

"No, Sir, I never have."

"Dost know nothing at all about whaling, I dare say—eh?"

"Nothing, Sir; but I have no doubt I shall soon learn. I've been several voyages in the merchant service, and I think that— "

"Marchant service be damned. Talk not that lingo to me. Dost see that leg?—I'll take that leg away from thy stern, if ever thou talkest of the marchant service to me again. Marchant service indeed! I suppose now ye feel considerable proud of having served in those marchant ships. But flukes! man, what makes thee want to go a whaling, eh?—it looks a little suspicious, don't it, eh?—Hast not been a pirate, hast thou?—Didst not rob thy last Captain, didst thou?—Dost not think of murdering the officers when thou gettest to sea?"

I protested my innocence of these things. I saw that under the mask of these half humorous innuendoes, this old sea-man, as an insulated Quakerish Nantucketer, was full of his insular prejudices, and rather distrustful of all aliens, unless they hailed from Cape Cod or the Vineyard.

"But what takes thee a-whaling? I want to know that before I think of shipping ye."

"Well, sir, I want to see what whaling is. I want to see the world."

"Want to see what whaling is, eh? Have ye clapped eye on Captain Ahab?"

"Who is Captain Ahab, sir?"

"Aye, aye, I thought so. Captain Ahab is the Captain of this ship."

"I am mistaken then. I thought I was speaking to the Captain himself."

"Thou art speaking to Captain Peleg—that's who ye are speaking to, young man. It belongs to me and Captain Bildad to see the Pequod fitted out for the voyage, and supplied with all her needs, including crew. We are part owners and agents. But as I was going to say, if thou wantest to know what whaling is, as thou tellest ye do, I can put ye in a way of finding it out before ye bind yourself to it, past backing out. Clap eye on Captain Ahab, young man, and thou wilt find that he has only one leg."

"What do you mean, sir? Was the other one lost by a whale?"

Going a-whalin', circa 1900. While a whaleship is fitted out and made ready for a voyage, men gather around what appears to be a makeshift sign-up "tent" made from an old wagon. — Joseph S. Martin photograph.

Kendall Collection, NBWM

"Lost by a whale! Young man, come nearer to me: it was devoured, chewed up, crunched by the monstrousest parmacetty that ever chipped a boat!—ah, ah!"

I was a little alarmed by his energy, perhaps also a little touched at the hearty grief in his concluding exclamation, but said as calmly as I could, "What you say is no doubt true enough, sir; but how could I know there was any peculiar ferocity in that particular whale, though indeed I might have inferred as much from the simple fact of the accident."

"I have given thee a hint about what whaling is; do ye yet feel inclined for it?"

"I do, sir."

"Now then, thou not only wantest to go a-whaling, to find out by experience what whaling is, but ye also want to go in order to see the world? Was not that what ye said? I thought so. Well then, just step forward there, and take a peep over the weather-bow, and then back to me and tell me what ye see there."

Going forward and glancing over the weather bow, I perceived that the ship swinging to her anchor with the flood-tide, was now obliquely pointing towards the open ocean. The prospect was unlimited, but exceedingly monotonous and forbidding; not the slightest variety that I could see.

"Well, what's the report?" said Peleg when I came back; "what did ye see?"

"Not much," I replied— "nothing but water; considerable horizon though, and there's a squall coming up, I think."

"Well, what dost thou think then of seeing the world? Can't ye see the world where you stand?"

I was a little staggered, but go a-whaling I must, and I would; and the Pequod was as good a ship as any—I thought the best—and all this I now repeated to Peleg. Seeing me so determined, he expressed his willingness to ship me.

"And thou mayest as well sign the papers right off," he added—"come along with ye." And so saying, he led the way below deck into the cabin.

Seated on the transom was what seemed to me a most uncommon and surprising figure. It turned out to be Captain Bildad, who along with Captain Peleg was one of the largest owners of the vessel; the other shares being held by a crowd of old annuitants; widows, fatherless children, and chancery wards; each owning about the value of a timber head, or a foot of plank, or a nail or two in the ship. People in Nantucket invest their money in whaling vessels, the same way that you do yours in approved state stocks bringing in good interest.

Now, Bildad, like Peleg, and indeed many other Nantucketers, was a Quaker, the island having been origi-

Quaker Captains, circa 1880. Thomas Nye and George Howland, Jr. (seated) survey their shipping interests. Howland (1806-92) was one of New Bedford's most successful whaling merchants. He also served as mayor during the Civil War. Although, like most Quakers, he opposed war, as a Unionist, abolitionist, and whaling merchant, he acquiesced. When draft riots broke out in the North, he garrisoned City Hall and stationed mounted guards at roads leading into town to keep out agitators. — *Joseph G. Tirrell photograph.*

nally settled by that sect; and to this day its inhabitants in general retain in an uncommon measure the peculiarities of the Quaker, only variously and anomalously modified by things altogether alien and heterogeneous. For some of these same Quakers are the most sanguinary of all sailors and whale-hunters. They are fighting Quakers; they are Quakers with a vengeance.

Like Captain Peleg, Captain Bildad was a well-to-do, retired whaleman. But unlike Captain Peleg—who cared not a rush for what are called serious things, and indeed deemed those selfsame serious things the veriest of all trifles—Captain Bildad had not only been originally educated according to the strictest sect of Nantucket Quakerism, but all his subsequent ocean life, and the sight of many unclad, lovely island creatures, round the Horn—all that had not moved this native born Quaker one single jot, had not so much as altered one angle of his vest. Still, for all this immutableness, was there some lack of common consistency about worthy Captain Bildad. Though refusing, from conscientious scruples, to bear arms against land invaders, yet himself had illimitably invaded the Atlantic and Pacific; and though a sworn foe to human bloodshed, yet had he in his straight-bodied coat, spilled tuns upon tuns of leviathan gore. How now in the contemplative evening of his days, the pious Bildad reconciled these things in the reminiscence, I do not know; but it did not seem to concern him much, and very probably he had long since come to the sage and sensible conclusion that a man's religion is one thing, and this practical world quite another.

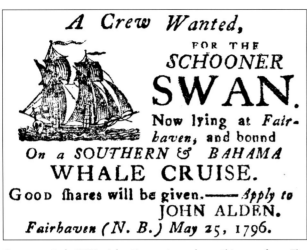

Crew wanted, 1796. *Advertisements such as this one from* The Medley, or New Bedford Marine Journal *appeared regularly throughout the 19th century in various newspapers published in southern New England.* — New Bedford Whaling Museum.

Now Bildad, I'm sorry to say, had the reputation of being an incorrigible old hunks, and in his sea-going days, a bitter, hard task-master. For a pious man, especially for a Quaker, he was certainly rather hard-hearted, to say the least. He never used to swear, though, at his men, they said; but somehow he got an inordinate quantity of cruel, unmitigated hard work out of them. Indolence and idleness perished from before him.

His own person was the exact embodiment of his utilitarian character. On his long, gaunt body, he carried no spare flesh, no superfluous beard, his chin having a soft, economical nap to it, like the worn nap of his broad-brimmed hat.

Such, then, was the person that I saw seated on the transom when I followed Captain Peleg down into the cabin. The space between the decks was small; and there, bolt-upright, sat old Bildad, who always sat so, and never leaned, and this to save his coat tails. His broad-brim was placed beside him; his legs were stiffly crossed; his drab vesture was buttoned up to his chin; and spectacles on nose, he seemed absorbed in reading from a ponderous volume.

"He says he's our man, Bildad," said Peleg, "he wants to ship."

"Dost thee?" said Bildad, in a hollow tone, and turning round to me.

"I dost," said I unconsciously, he was so intense a Quaker.

"What do ye think of him, Bildad?" said Peleg.

"He'll do," said Bildad, eyeing me, and then went on spelling away at his book in a mumbling tone quite audible.

I thought him the queerest old Quaker I ever saw, especially as Peleg, his friend and old shipmate, seemed such a blusterer. But I said nothing, only looking round me sharply. Peleg now threw open a chest, and drawing forth the ship's articles, placed pen and ink before him, and seated himself at a little table. I began to think it was high time to settle with myself at what terms I would be willing to engage for the voyage. I was already aware that in the whaling business they paid no wages; but all hands, including the captain, received certain shares of the profits called *lays*, and that these lays were proportioned to the degree of importance pertaining to the respective duties of the ship's company. I was also aware that being a green hand at whaling, my own lay would not be very large; but considering that I was used to the sea, could steer a ship, splice a rope, and all that, I made no doubt that from all I had heard I should be offered at least the 275th lay—that is, the 275th part of the clear net proceeds of the voyage, whatever that might eventually amount to. And though the 275th lay was what they call a rather *long lay*, yet it was better than nothing; and if we had a lucky voyage, might pretty nearly pay for the clothing I would wear out on it, not to speak of my three years' beef and board, for which I would not have to pay.

It might be thought that this was a poor way to accumulate a princely fortune—and so it was, a very poor way indeed. But I am one of those that never take on about princely fortunes, and am quite content if the world is ready to board and lodge me, while I am putting up at this grim sign of the Thunder Cloud. Upon the whole, I thought that the 275th lay would be about the fair thing, but would not have been surprised had I been offered the 200th, considering I was of a broad-shouldered make.

But one thing, nevertheless, that made me a little distrustful about receiving a generous share of the profits was this: Ashore, I had heard something of both Captain Peleg and his unaccountable old crony Bildad; how that they being the principal proprietors of the Pequod, therefore the other and more inconsiderable and scattered owners, left nearly the whole management of the ship's affairs to these two. And I did not know but what the stingy old Bildad might have a mighty deal to say about shipping hands, especially as I now found him on board the Pequod, quite at home there in the cabin, and reading his Bible as if at his own fireside. Bildad never heeded us, but went on mumbling to himself out of his book, "Lay not up for yourselves treasures upon earth, where moth—'"

"Well, Captain Bildad," interrupted Peleg, "what d'ye say, what lay shall we give this young man?"

"Thou knowest best," was the sepulchral reply, "the seven hundred and seventy-seventh wouldn't be too much, would it?— 'where moth and rust do corrupt, but lay—'"

"*Lay*, indeed," thought I, and such a lay! It was an exceedingly long lay that, indeed; and though from the magnitude of the figure it might at first deceive a

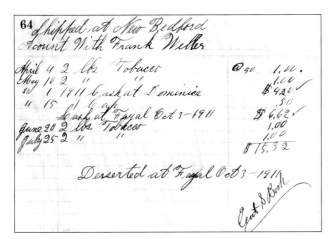

Derelict account, 1911. Frank Webber won't be paying this account for tobacco—he jumped ship in Fayal. — New Bedford Free Public Library.

landsman, yet the slightest consideration will show that though seven hundred and seventy-seven is a pretty large number, yet, when you come to make a tenth of it, you will then see, I say, that the seven hundred and seventy-seventh part of a farthing is a good deal less than seven hundred and seventy-seven gold doubloons; and so I thought at the time.

"Why, blast your eyes, Bildad," cried Peleg, "Thou dost not want to swindle this young man! he must have more than that."

"Seven hundred and seventy-seventh," again said Bildad, without lifting his eyes; and then went on mumbling— " 'for where your treasure is, there will your heart be also.' "

"I am going to put him down for the three hundredth," said Peleg, "do ye hear that, Bildad! The three hundredth lay, I say."

Bildad laid down his book, and turning solemnly towards him said, "Captain Peleg, thou hast a generous heart; but thou must consider the duty thou owest to the other owners of this ship—widows and orphans, many of them—and that if we too abundantly reward the labors of this young man, we may be taking the bread from those widows and those orphans. The seven hundred and seventy-seventh lay, Captain Peleg."

"Thou Bildad!" roared Peleg, starting up and clattering about the cabin. "Blast ye, Captain Bildad, if I had followed thy advice in these matters, I would afore now had a conscience to lug about that would be heavy enough to founder the largest ship that ever sailed round Cape Horn."

"Captain Peleg," said Bildad steadily, "thy conscience may be drawing ten inches of water, or ten fathoms, I can't tell; but as thou art still an impenitent man, Captain Peleg, I greatly fear lest thy conscience be but a leaky one; and will in the end sink thee foundering down to the fiery pit, Captain Peleg."

"Fiery pit! fiery pit! ye insult me, man; past all natural bearing, ye insult me. It's an all-fired outrage to tell any human creature that he's bound to hell. Flukes and flames! Bildad, say that again to me, and start my soul-bolts, but I'll—I'll—yes, I'll swallow a live goat with all his hair and horns on. Out of the cabin, ye canting, drab-colored son of a wooden gun—a straight wake with ye!"

Alarmed at this terrible outburst between the two principal and responsible owners of the ship, I stepped aside from the door to give egress to Bildad, who, I made no doubt, was all eagerness to vanish from before the awakened wrath of Peleg. But to my astonishment, he sat down again on the transom very quietly, and seemed to have not the slightest intention of withdrawing. As for Peleg, after letting off his rage as he had, there seemed no more left in him, and he, too, sat down like a lamb, though he twitched a little as if still nervously agitated. "Now then, my young man, Ishmael's thy name, didn't ye say? Well then, down ye go here, Ishmael, for the three hundredth lay."

"Captain Peleg," said I, "I have a friend with me who wants to ship too."

Balance of payment, 1912. A page taken from the crew account book of the Alice Knowles *in 1911 shows typical deductions from a whaleman's earnings. This man returned to the Dominican Republic with $63.00 for a year of whaling.*

New Bedford Free Public Library

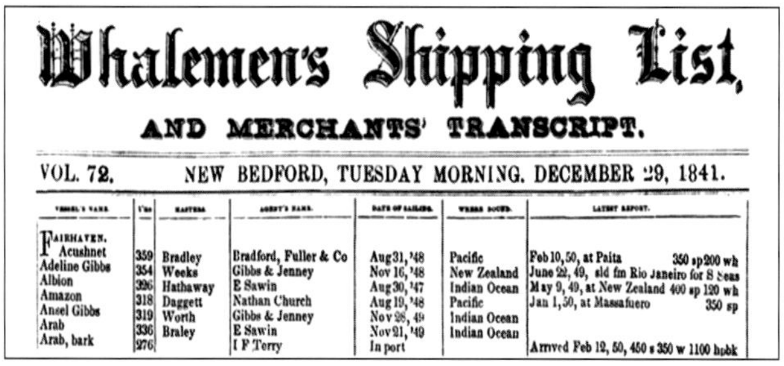

"Whalemen's Shipping List." This well-known tabloid provided useful data on American whaleships. Information on the Acushnet shown here was for her final voyage. — Digitally re-composed from two newspaper volumes.

"Fetch him along, and we'll look at him."

"What lay does he want?" groaned Bildad, glancing up from the book in which he had again been burying himself.

"Oh! never thee mind about that, Bildad," said Peleg. "Has he ever whaled it any?" turning to me.

"Killed more whales than I can count, Captain Peleg."

"Well, bring him along then." And, after signing the papers, off I went; nothing doubting but that I had done a good morning's work, and that the Pequod was the identical ship that Yojo had provided to carry Queequeg and me round the Cape.

But I had not proceeded far, when I began to bethink me that the captain with whom I was to sail yet remained unseen by me; it is always as well to have a look at him before irrevocably committing yourself into his hands. Turning back I accosted Captain Peleg, inquiring where Captain Ahab was to be found.

"And what dost thou want of Captain Ahab? It's all right enough; thou art shipped."

"Yes, but I should like to see him."

"But I don't think thou wilt be able to at present. I don't know exactly what's the matter with him; a sort of sick, and yet he don't look so. In fact, he ain't sick; but no, he isn't well either. He's a queer man, Captain Ahab—so some think—but a good one. Oh, thou'lt like him well enough; no fear, no fear. He's a grand, ungodly, god-like man, Captain Ahab; doesn't speak much; but, when he does speak, then you may well listen. Mark ye, be forewarned; Ahab's above the common; Ahab's been in colleges, as well as 'mong the cannibals; been used to deeper wonders than the waves; fixed his fiery lance in mightier, stranger foes than whales. His lance! aye, the keenest and the surest that out of all our isle! Oh! he ain't Captain Bildad; no, and he ain't Captain Peleg; *he's Ahab*, boy.

"I know Captain Ahab well; I've sailed with him as mate years ago; I know what he is—a good man—not a pious, good man, like Bildad, but a swearing good man—something like me—only there's a good deal more of him. I know, too, that ever since he lost his leg last voyage by that accursed whale, he's been a kind of moody—desperate moody, and savage sometimes; but that will all pass off. And once for all, let me tell thee and assure thee, young man, it's better to sail with a moody good captain than a laughing bad one. Besides, my boy, he has a wife—not three voyages wedded—a sweet, resigned girl. Think of that; by that sweet girl that old man has a child: hold ye then there can be any utter, hopeless harm in Ahab? No, no, my lad; stricken, blasted, if he be, Ahab has his humanities!"

As I walked away, I was full of thoughtfulness; what had been incidentally revealed to me of Captain Ahab, filled me with a certain wild vagueness of painfulness concerning him. And somehow, at the time, I felt a sympathy and a sorrow for him, but for I don't know what, unless it was the cruel loss of his leg. And yet I also felt a strange awe of him; but that sort of awe, which I cannot at all describe, was not exactly awe; I do not know what it was. However, my thoughts were at length carried in other directions, so that for the present dark Ahab slipped my mind.

Detail from "Shipping Paper" for the crew of the Acushnet, 1840. *Similar to the Shipping List at right, this document in the collection of the New Bedford Free Public Library gives additional information, such as the date the sailor signed up, the quality of the seaman (such as "Green Hand" or "Boat Steerer"), and the "lay" or pay he would receive. Melville (second from bottom) signed on as a green hand on December 25 and was scheduled to receive a one-hundred-forty-fifth lay. His friend, Toby Greene, is second from top. Had they completed their voyage rather than overstay their visit in Polynesia, the young comrades would have returned home with about $200 (less expenses owed the ship) for three years' whaling.*

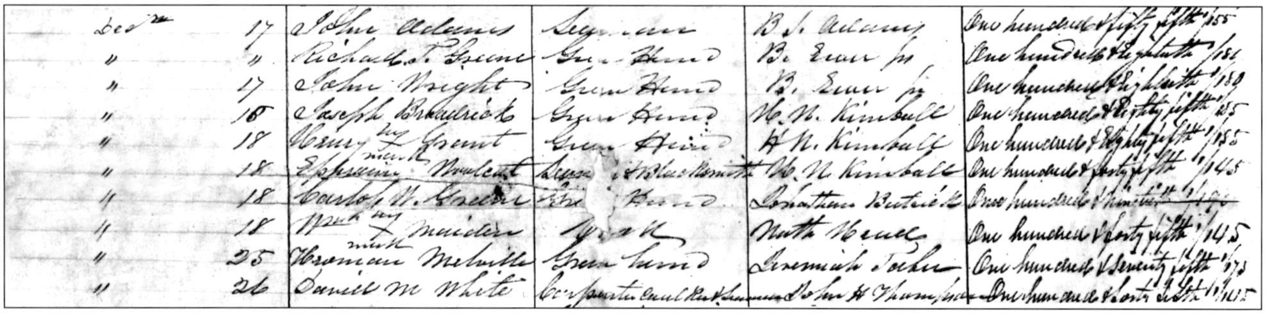

New Bedford Free Public Library

Shipping Paper from the Acushnet, 1840. A second document or "shipping paper" listing information about the crew of the Acushnet. Here, sailors listed place of birth, place of residence, country of citizenship, age, height, complexion, and hair color. Interestingly, although most of them claim local residency (Fairhaven-11, New Bedford-9, Nantucket-2, Rochester-1), only one of the 25, a Nantucketer, claims a local birthplace. Among the crew are 21 Americans, two Portuguese, one Cape Verdean and one Englishman. Two have black complexions, one mulatto, 11 dark-skinned and 11 light-skinned. All are between 5'2" and 5'11" except one—a 6'6" 18-year-old with dark skin and dark hair named Joseph Broadrick (possibly Cape Verdean) who listed Boston as his place of birth. Could his height be an error in writing, or is this the giant who inspired the character Daggoo?

The document also contains contractual information: In fine print (not shown), consignees are told "to do their duty and obey the law… as become good and faithful Seamen or Mariners, while cruising for whales…." If those responsible failed to enforce regulations, such as keeping women and alcohol off the ship, they would forfeit 20 days' pay.

New Bedford Whaling Museum

Chapter XVIII

His Mark

As we were walking down the end of the wharf towards the ship, Queequeg carrying his harpoon, Captain Peleg in his gruff voice loudly hailed us from his wigwam, saying he had not suspected my friend was a cannibal, and furthermore announcing that he let no cannibals on board that craft, unless they previously produced their papers.

"What do you mean by that, Captain Peleg?" said I, now jumping on the bulwarks, and leaving my comrade standing on the wharf.

"I mean," he replied, "he must show his papers."

"Yea," said Captain Bildad in his hollow voice, sticking his head from behind Peleg's, out of the wigwam. "He must show that he's converted. Son of darkness," he added, turning to Queequeg, "art thou at present in communion with any christian church?"

"Why," said I, "he's a member of the First Congregational Church." Here be it said, that many tattooed savages sailing in Nantucket ships at last come to be converted into the churches.

"First Congregational Church," cried Bildad, "what! that worships in Deacon Deuteronomy Coleman's meeting-house?" and so saying, taking out his spectacles, he rubbed them with his great yellow bandana handkerchief, and putting them on very carefully, came out of the wigwam, and leaning stiffly over the bulwarks, took a good long look at Queequeg.

"Young man," said Bildad sternly, "thou art skylarking with me—explain thyself, thou young Hittite. What church dost thee mean? answer me."

Finding myself thus hard pushed, I replied. "I mean, sir, the same ancient Catholic Church to which you and I, and Captain Peleg there, and Queequeg here, and all of us, and every mother's son and soul of us belong; the great and everlasting First Congregation of this whole worshipping world; we all belong to that; only some of us cherish some queer crotchets noways touching the grand belief; in that we all join hands."

"Splice, thou mean'st splice hands," cried Peleg, drawing nearer. "Young man, you'd better ship for a missionary, instead of a fore-mast hand; I never heard a better sermon. Come aboard, come aboard; never mind about the papers. I say, tell Quohog there—what's that you call him? tell Quohog to step along. By the great anchor, what a harpoon he's got there! looks like good stuff that; and he handles it about right. I say, Quohog,

Harpooners on the *John R. Manta* strap on their irons, as they prepare for voyage, 1925. — *William R. Hegarty Collection.*

or whatever your name is, did you ever stand in the head of a whale-boat? did you ever strike a fish?"

Without saying a word, Queequeg, in his wild sort of way, jumped upon the bulwarks, from thence into the bows of one of the whale-boats hanging to the side; and then bracing his left knee, and poising his harpoon, cried out in some such way as this:—

"Cap'ain, you see him small drop tar on water dere? You see him? well, spose him one whale eye, well, den!" and taking sharp aim at it, he darted the iron right over old Bildad's broad brim, clean across the ship's decks, and struck the glistening tar spot out of sight.

"Now," said Queequeg, quietly hauling in the line, "spos-ee him whale-e eye; why, dad whale dead."

"Quick, Bildad," said Peleg, his partner, who, aghast at the close vicinity of the flying harpoon, had retreated towards the cabin gangway. "Quick, I say, you Bildad, and get the ship's papers. We must have Hedgehog there, I mean Quohog, in one of our boats. Look ye, Quohog, we'll give ye the ninetieth lay, and that's more than ever was given a harpooneer yet out of Nantucket."

So down we went into the cabin, and to my great joy Queequeg was soon enrolled among the same ship's company to which I myself belonged.

When all preliminaries were over and Peleg had got everything ready for signing, he turned to me and said, "I guess Quohog there don't know how to write, does he? I say, Quohog, blast ye! dost thou sign thy name or make thy mark?"

But at this question, Queequeg, who had twice or thrice before taken part in similar ceremonies, looked no ways abashed; but taking the offered pen, copied upon the paper, in the proper place, an exact counterpart of a queer round figure which was tattooed upon his arm; so that through Captain Peleg's obstinate mistake touching his appellative, it stood something like this:

Chapter xx

All Astir

A day or two passed, and there was great activity aboard the Pequod. Not only were the old sails being mended, but new sails were coming on board, and bolts of canvas, and coils of rigging; in short, everything betokened that the ship's preparations were hurrying to a close.

Frank Lewis—shipkeeper, stevedore, former whaleman and longtime fixture on the New Bedford waterfront—makes sure the ship is fitted out properly, circa 1910. — *William R. Hegarty Collection.*

On the day following Queequeg's signing the articles, word was given at all the inns where the ship's company were stopping, that their chests must be on board before night, for there was no telling how soon the vessel might be sailing. So Queequeg and I got down our traps, resolving, however, to sleep ashore till the last. But it seems they always give very long notice in these cases, and the ship did not sail for several days. But no wonder; there was a good deal to be done, and there is no telling how many things to be thought of, before the Pequod was fully equipped.

Every one knows what a multitude of things—beds, sauce-pans, knives and forks, shovels and tongs, napkins, nut-crackers, and what not, are indispensable to the business of housekeeping. Just so with whaling, which necessitates a three-years' housekeeping upon the wide ocean, far from all grocers, costermongers, doctors, bakers, and bankers. And though this also holds true of merchant vessels, yet not by any means to the same extent as with whale men. For besides the great length of the whaling voyage, the numerous articles peculiar to the prosecution of the fishery, and the impossibility of replacing them at the remote harbors usually frequented, it must be remembered, that of all ships, whaling vessels are the most exposed to accidents of all kinds, and especially to the destruction and loss of the very things upon which the success of the voyage most depends. Hence, the spare boats, spare spars, and spare lines and harpoons, and spare everythings, almost, but a spare captain and duplicate ship.

Fitting-out the Wanderer, New Bedford, 1918. *"As for Bildad, he carried about with him a long list of the articles needed, and at every fresh arrival, down went his mark opposite that article upon the paper. Every once in a while Peleg came hobbling out of his whalebone den, roaring at the men down the hatchways, roaring up to the riggers at the mast-head, and then concluded by roaring back into his wigwam."* — *Chapter 20. Arthur F. Packard photograph.*

Courtesy of T. E. Holcombe

At the period of our arrival at the Island, the heaviest storage of the Pequod had been almost completed; comprising her beef, bread, water, fuel, and iron hoops and staves. But, for some time there was a continual fetching and carrying on board of divers odds and ends of things, both large and small.

Every once and a while Peleg came hobbling out of his whalebone den, roaring at the men down the hatchways, roaring up to the riggers at the mast-head, and then concluded by roaring back into his wigwam.

During these days of preparation, Queequeg and I often visited the craft, and as often I asked about Captain Ahab, and how he was, and when he was going to come on board his ship. To these questions they would answer, that he was getting better and better, and was expected aboard every day; meantime, the two Captains, Peleg and Bildad, could attend to everything necessary to fit the vessel for the voyage.

At last it was given out that some time next day the ship would certainly sail. So next morning, Queequeg and I took a very early start.

Shipping out, 1918. *Crew of the bark* Greyhound *load their slop chests and personal belongings for what would be the ship's last whaling voyage.* — Clifford W. Ashley photograph. New Bedford Whaling Museum.

Unloading a whaleship with a donkey engine, 1859. *Artist David Hunter Strother, a native of West Virginia and a soldier with the Union Army, made a series of sketches accurately detailing work on the New Bedford waterfront. Steam-powered donkey engines helped stevedores load and unload casks weighing as much as a half-ton each.*

Kendall Collection, NBWM

Chapter XXI

Going Aboard

It was nearly six o'clock, but only grey imperfect misty dawn, when we drew nigh the wharf.

"There are some sailors running ahead there, if I see right," said I to Queequeg, 'it can't be shadows; she's off by sunrise, I guess; come on!"

Stepping on board the Pequod, we found everything in profound quiet, not a soul moving. The cabin entrance was locked within; the hatches were all on, and lumbered with coils of rigging. Going forward to the forecastle, we found the slide of the scuttle open. Seeing a light, we went down, and found only an old rigger there, wrapped in a tattered pea-jacket. He was thrown at whole length upon two chests, his face downwards and inclosed in his folded arms. The profoundest slumber slept upon him.

"Those sailors we saw, Queequeg, where can they have gone to?" said I, looking dubiously at the sleeper. Queequeg put his hand upon the sleeper's rear, as though feeling if it was soft enough; and then, without more ado, sat quietly down there.

"Gracious! Queequeg, don't sit there," said I.

Queequeg removed himself to just beyond the head of the sleeper, and lighted his tomahawk pipe. I sat at the feet. We kept the pipe passing over the sleeper, from one to the other. Meanwhile, upon questioning him in his broken fashion, Queequeg gave me to understand that, in his land, owing to the absence of settees and sofas of all sorts, the king, chiefs, and great people generally, were in the custom of fattening some of the lower orders for ottomans; and to furnish a house comfortably in that respect, you had only to buy up eight or ten lazy fellows, and lay them round in the piers and alcoves. Besides, it was very convenient on an excursion; much better than those garden-chairs which are convertible into walking-sticks; upon occasion, a chief calling his attendant, and desiring him to make a settee of himself under a spreading tree, perhaps in some damp marshy place.

While narrating these things, every time Queequeg received the tomahawk from me, he flourished the hatchet-side of it over the sleeper's head.

"What's that for, Queequeg?"

"Perry easy, kill-e; oh! perry easy!"

He was going on with some wild reminiscences about his tomahawk-pipe, which, it seemed, had in its two uses both brained his foes and soothed his soul, when we were directly attracted to the sleeping rigger. The strong vapor now completely filling the contracted hole, it began to tell

Tying up sheets on the topmast, 1920.
— *Pardon Gifford photograph. Kendall Collection, NBWM.*

upon him. He breathed with a sort of muffledness; then seemed troubled in the nose; then revolved over once or twice; then sat up and rubbed his eyes.

"Holloa!" he breathed at last, "who be ye smokers?"

"Shipped men," answered I, "when does she sail?"

"Aye, aye, ye are going in her, be ye? She sails to-day. The Captain came aboard last night."

"What Captain?—Ahab?"

"Who but him indeed?"

I was going to ask him some further questions concerning Ahab, when we heard a noise on deck.

"Holloa! Starbuck's astir," said the rigger. "He's a lively chief mate, that; good man, and a pious; but all alive now, I must turn to." And so saying he went on deck, and we followed.

It was now clear sunrise. Soon the crew came on board in twos and threes; the riggers bestirred themselves; the mates were actively engaged; and several of the shore people were busy in bringing various last things on board. Meanwhile Captain Ahab remained invisibly enshrined.

On the foredeck of the* Wanderer, *1918. *Taking up chain on the windlass, the crew makes final preparations. The ship is then towed into the "stream" where she is boarded by her owners, the crew, and families and friends. Later, all will take leave of the crew and bid farewell as the ship, under tow, leaves the harbor.* — *Arthur F. Packard photograph.*

Courtesy of T. E. Holcombe

Chapter XXII

Merry Christmas

At length, towards noon, upon the final dismissal of the ship's riggers, and after the Pequod had been hauled out from the wharf, the two captains, Peleg and Bildad, issued from the cabin, and turning to the chief mate, Peleg said:

"Now, Mr. Starbuck, are you sure everything is right? Captain Ahab is all ready—just spoke to him—nothing more to be got from shore, eh? Well, call all hands, then. Muster 'em aft here—blast 'em!"

"No need of profane words, however great the hurry, Peleg," said Bildad, "but away with thee, friend Starbuck, and do our bidding."

As for Captain Ahab, no sign of him was yet to be seen; only, they said he was in the cabin. But then, the idea was, that his presence was by no means necessary in getting the ship under weigh, and steering her well out to sea. Indeed, as that was not at all his proper business, but

"Bound Out," circa 1860. — *Watercolor by Benjamin Russell.*

the pilot's; and as he was not yet completely recovered—so they said—therefore, Captain Ahab stayed below.

"Strike the tent there!"—was the next order. As I hinted before, this whalebone marquee was never pitched except in port; and on board the Pequod, for thirty years, the order to strike the tent was well known to be the next thing to heaving up the anchor.

"Man the capstan! Blood and thunder!—jump!"—was the next command, and the crew sprang for the handspikes.

Now, in getting under weigh, the station generally occupied by the pilot is the forward part of the ship.

Bark Swallow, *circa 1880.* Her anchor broken (pulled up) and her crew all aboard, the ship is towed to the outer harbor where, after clearing the headlands, she will loosen her sails and head into the Atlantic. Swallow *was lost at sea in 1901*. — *Joseph G. Tirrell photograph.*

Spinner Collection

And here Bildad might now be seen actively engaged in looking over the bows for the approaching anchor, and at intervals singing what seemed a dismal stave of psalmody, to cheer the hands at the windlass, who roared forth some sort of a chorus about the girls in Boople Alley, with hearty good will. Nevertheless, not three days previous, Bildad had told them that no profane songs would be allowed on board the Pequod, particularly in getting under weigh; and Charity, his sister, had placed a small choice copy of Watts in each seaman's berth.

At last the anchor was up, the sails were set, and off we glided. It was a short, cold Christmas; and as the short northern day merged into night, we found ourselves almost broad upon the wintry ocean, whose freezing spray cased us in ice, as in polished armor. The long rows of teeth on the bulwarks glistened in the moonlight; and like the white ivory tusks of some huge elephant, vast curving icicles depended from the bows.

Spite of this frigid winter night in the boisterous Atlantic, spite of my wet feet and wetter jacket, there was yet, it then seemed to me, many a pleasant haven in store; and meads and glades so eternally vernal, that the grass shot up by the spring, untrodden, unwilted, remains at midsummer.

At last we gained such an offing, that the two pilots were needed no longer. The stout sail-boat that had accompanied us began ranging alongside.

Peleg turned to his comrade, with a final sort of look about him,—"Captain Bildad—come, old shipmate, we must go. Back the main-yard there! Boat ahoy! Stand by to come close alongside, now! Careful, careful!—come, Bildad, boy—say your last. Luck to ye, Starbuck—luck to ye, Mr. Stubb—luck to ye, Mr. Flask—good-bye, and good luck to ye all—and this day three years I'll have a hot supper smoking for ye in old Nantucket. Hurrah and away!"

"God bless ye, and have ye in His holy keeping, men," murmured old Bildad, almost incoherently. "I hope ye'll have fine weather now, so that Captain Ahab may soon be moving among ye—a pleasant sun is all he needs, and ye'll have plenty of them in the tropic voyage ye go. Be careful in the hunt, ye mates. Don't stave the boats needlessly, ye harpooneers; good white cedar plank is raised full three per cent. within the year. Don't forget your prayers, either. Mr Starbuck, mind that cooper don't waste the spare staves. Oh! the sail-needles are in the green locker! Don't whale it too much a' Lord's days, men; but don't miss a fair chance either, that's rejecting Heaven's good gifts. Have an eye to the molasses tierce, Mr. Stubb; it was a little leaky, I thought. If ye touch at the islands, Mr. Flask, beware of fornication. Good-bye, good-bye! Don't keep that cheese too long down in the hold, Mr. Starbuck; it'll spoil. Be careful with the butter—twenty cents the pound it was, and mind ye, if—"

"Come, come, Captain Bildad; stop palavering,—away!" and with that, Peleg hurried him over the side, and both dropt into the boat.

Ship and boat diverged; the cold, damp night breeze blew between; a screaming gull flew overhead; the two hulls wildly rolled; we gave three heavy-hearted cheers, and blindly plunged like fate into the lone Atlantic.

"Outward Bound," 1922. *The* Charles W. Morgan *leaves a lightship in her wake, as she heads full sail into the open seas.*

"Sweet fields beyond the swelling flood,
Stand dressed in living green.
So to the Jews old Canaan stood,
While Jordan rolled between."

"Never did those sweet words sound more sweetly to me than then. They were full of hope and fruition. Spite of this frigid winter night in the boisterous Atlantic, spite of my wet feet and wetter jacket, there was yet, it then seemed to me, many a pleasant haven in store; and meads and glades so eternally vernal, that the grass shot up by the spring, untrodden, unwilted, remains at midsummer." — *Chapter 22. Etching by J. Duncan Gleason.*

William R. Hegarty Collection

Chapter XXIII, XXIV

The Lee Shore · The Advocate

Some chapters back, one Bulkington was spoken of, a tall, new-landed mariner, encountered in New Bedford at the inn.

When on that shivering winter's night, the Pequod thrust her vindictive bows into the cold malicious waves, who should I see standing at her helm but Bulkington! I looked with sympathetic awe and fearfulness upon the man, who in mid-winter just landed from a four years' dangerous voyage, could so unrestingly push off again for still another tempestuous term. The land seemed scorching to his feet. The port would fain give succor. But in that gale, the port, the land, is that ship's direst jeopardy; she must fly all hospitality.

CHAPTER XXIV

Doubtless one leading reason why the world declines honoring us whalemen, is this: they think that, at best, our vocation amounts to a butchering sort of business; and that when actively engaged therein, we are surrounded by all manner of defilements. Butchers we are, that is true. But butchers, also, and butchers of the bloodiest badge have been all Martial Commanders whom the world invariably delights to honor.

But, though the world scouts at us whale hunters, yet does it unwittingly pay us the profoundest homage; yea, an all-abounding adoration! for almost all the tapers, lamps, and candles that burn round the globe, burn, as before so many shrines, to our glory!

How comes it that we whalemen of America now outnumber all the rest of the banded whalemen in the world; sail a navy of upwards of seven hundred vessels; manned by eighteen thousand men; yearly consuming 4,000,000 of dollars; the ships worth, at the time of sailing, $20,000,000; and every year importing into our harbors a well reaped harvest of $7,000,000? How comes all this, if there be not something puissant in whaling?

But this is not the half; look again.

For many years past the whale-ship has been the pioneer in ferreting out the remotest and least known parts of the earth. She has explored seas and archipelagoes which had no chart, where no Cook or Vancouver had ever sailed. If American and European men-of-war now peacefully ride in once savage harbors, let them fire salutes to the honor and glory of the whale-ship, which originally showed them the way, and first interpreted between them and the savages. They may celebrate as they will the heroes of Exploring Expeditions, your Cookes, your Krusensterns; but I say that scores of anonymous Captains have sailed out of Nantucket, that were as great, and greater than your Cooke and your Krusenstern. For in their succorless empty-handedness, they, in the heathenish sharked waters, and by the beaches of unrecorded, javelin islands, battled with virgin wonders and terrors that Cooke with all his marines and muskets would not will-

Harper's Magazine's 1860 frontispiece of a writer's travelogue through New Bedford romanticizing this "butchering sort of business."

Bark Lancer off Cape Thaddeus, Siberia. *The first American whaler passed through the Bering Strait in 1819 and by the 1820s American whalers were exploring the most remote seas on earth in search of whaling grounds. The search became especially urgent in the 1840s and 50s as the sperm whale population was depleted. The first bowhead whales were taken by New Bedford whalers in the North Pacific in the mid 1840s, and it wasn't long before other whalers realized the richness of the Arctic whaling grounds. By the 1860s and 70s the Arctic Region was the principal whaling grounds of the New England fleet. No waters on earth were too harsh or remote for American whalers.* —Herbert L. Aldrich photograph.

New Bedford Whaling Museum

ingly have dared. All that is made such a flourish of in the old South Sea Voyages, those things were but the lifetime commonplaces of our heroic Nantucketers. Often, adventures which Vancouver dedicates three chapters to, these men accounted unworthy of being set down in the ship's common log. Ah, the world! Oh, the world!

Until the whale fishery rounded Cape Horn, no commerce but colonial, scarcely any intercourse but colonial, was carried on between Europe and the long line of the opulent Spanish provinces on the Pacific coast. It was the whaleman who first broke through the jealous policy of the Spanish crown, touching those colonies; and, if space permitted, it might be distinctly shown how from those whalemen at last eventuated the liberation of Peru, Chili, and Bolivia from the yoke of Old Spain, and the establishment of the eternal democracy in those parts.

That great America on the other side of the sphere, Australia, was given to the enlightened world by the whaleman. The whale-ship is the true mother of that now mighty colony. Moreover, in the infancy of the first Australian settlement, the emigrants were several times saved from starvation by the benevolent biscuit of the whale-ship luckily dropping an anchor in their waters. If that double-bolted land, Japan, is ever to become hospitable, it is the whale-ship alone to whom the credit will be due; for already she is on the threshold.

And, as for me, if, by any possibility, there be any as yet undiscovered prime thing in me; if I shall ever deserve any real repute in that small but high hushed world which I might not be unreasonably ambitious of; if hereafter I shall do anything that, upon the whole, a man might rather have done than to have left undone; if, at my death, my executors, or more properly my creditors, find any precious MSS. in my desk, then here I prospectively ascribe all the honor and the glory to whaling; for a whale-ship was my Yale College and my Harvard.

Whaleship John J. Howland, 1841. How prophetic Melville's words: "If that double-bolted land, Japan, is ever to become hospitable, it is the whaleship alone to whom the credit will be due…" While Melville was dallying in the South Pacific, Captain William Whitfield of Fairhaven and his New Bedford whaleship were rescuing from a deserted isle in the Japan Sea five Japanese castaways. One of them was 14-year-old John Manjiro, whom Whitfield brought home to Fairhaven and helped raise and educate. Ten years later, as Moby-Dick was being published in London, Manjiro returned to Japan. He was immediately arrested for espionage and interrogated for nine months. His knowledge, however, would free him—and he became instrumental in the process that led to Japan beginning relations with the Western world. The painting of the John Howland was made by his interrogators (with Manjiro's help). An accomplished whaler, seaman and navigator, Manjiro went on to become a respected Samurai educator in Japan, teaching English, navigation, shipbuilding and other Western sciences learned in Fairhaven and at sea. — *Watercolor by Shoryo Kawada:* **Hyoson Kirryaku.**

Rosenbach Museum, Philadelphia

Knights and Squires · Knights and Squires

The chief mate of the Pequod was Starbuck, a native of Nantucket, and a Quaker by descent. He was a long, earnest man, and though born on an icy coast, seemed well adapted to endure hot latitudes, his flesh being hard as twice-baked biscuit. Looking into his eyes, you seemed to see there the yet lingering images of those thousand-fold perils he had calmly confronted through life. "I will have no man in my boat," said Starbuck, "who is not afraid of a whale." By this, he seemed to mean, not only that the most reliable and useful courage was that which arises from the fair estimation of the encountered peril, but that an utterly fearless man is a far more dangerous comrade than a coward.

CHAPTER XXVII

Stubb was the second mate. He was a native of Cape Cod; and hence, according to local usage, was called a Cape-Cod-man. A happy-go-lucky; neither craven nor valiant; taking perils as they came with an indifferent air; and while engaged in the most imminent crisis of the chase, toiling away, calm and collected as a journeyman joiner engaged for the year. Good-humored, easy, and careless, he presided over his whale-boat as if the most deadly encounter were but a dinner, and his crew all invited guests.

The third mate was Flask, a native of Tisbury, in Martha's Vineyard. A short, stout, ruddy young fellow, very pugnacious concerning whales, who somehow seemed to think that the great Leviathans had personally and hereditarily affronted him; and therefore it was a sort of point of honor with him, to destroy them whenever encountered.

Now these three mates—Starbuck, Stubb, and Flask, were momentous men. They it was who by universal prescription commanded three of the Pequod's boats as headsmen.

And since in this famous fishery, each mate or headsman, like a Gothic Knight of old, is always accompanied by his boat-steerer or harpooner; it is therefore but meet, that in this place we set down who the

Crew and guests on deck of the Wanderer, August 25, 1924. *The day after this photograph was taken, the Wanderer set sail. Captain Antone F. Edwards (center), recognizing the severity of the high winds, decided to drop anchor, wait out the storm and look for additional crew. Unfortunately, the storm became the Hurricane of 1924 and the Wanderer dragged anchor and ran aground on Sow and Pig Reef off Cuttyhunk Island. This incident marked the end of New Bedford's whaling industry, as no other square-rigged vessel would ever again go whaling from this port.*

Kendall Collection, NBWM

New Bedford Whaling Museum

New Bedford Whaling Museum

William R. Hegarty Collection

New Bedford Free Public Library

Private Collection

New Bedford Free Public Library

Faces of whaling men, 1850–1920, (clockwise from top left):

Jamie L. McKenzie, 1856. At 14, Jamie made his first whaling voyage on the bark Reindeer. By 23, he was first officer of the merchant vessel Simoda, but in January 1863, he was washed overboard and lost.

Amos Haskins, whaleship captain, circa 1850. A Wampanoag Indian, Amos sailed on six whaling voyages between 1839 and 1861. In 1851, he was appointed master of the bark Massasoit, a notable accomplishment for a Native American. He was lost at sea in 1861.

Quinton Degrasse, "cabin boy," 1917. Degrasse was one of two survivors of the bark Alice Knowles, which sank off Cape Hatteras, September 1917. Adrift at sea in a submerged and shattered whaleboat for four days, he and his friend Jules were picked up by a British warship. "To pull us in they dragged us [about a mile]. Since we couldn't move, we had to have ropes put around us and then were hoisted aboard ship. We were so crippled and frozen that they had to pry us apart." Jules died shortly after the rescue.

Whaling captains in the Arctic, 1887. Herbert Aldrich sailed into the Arctic aboard Young Phoenix and spent 87 days recording the experiences of the whaling fleet. "32 vessels were in the fleet this year," he wrote. "One of the very few years when every vessel that sailed [to the Arctic] returned to port."

Joe Bent, boatsteerer, Alice Knowles, **1914.** Joe was lost in the 1917 shipwreck (see above).

Captain Joseph F. Edwards (1886-1933). The youngest of three Azorean brothers who became whaling captains, Edwards took his first command on the Charles W. Morgan. Azorean whalers settled in Nantucket and Martha's Vineyard as early as the mid 1700s, and built a large neighborhood community in New Bedford by the early 1800s. Most were from Fayal, Flores or Pico, and many went on to become whaling masters.

Pequod's harpooneers were, and to what headsman each of them belonged.

First of all was Queequeg, whom Starbuck, the chief mate, had selected for his squire. But Queequeg is already known.

Next was Tashtego, an unmixed Indian from Gay Head, the most westerly promontory of Martha's Vineyard, where there still exists the last remnant of a village of red men, which has long supplied the neighboring island of Nantucket with many of her most daring harpooneers. In the fishery, they usually go by the generic name of Gay-Headers. Tashtego's long, lean, sable hair, his high cheek bones, and black rounding eyes—for an Indian, Oriental in their largeness, but Antarctic in their glittering expression—all this sufficiently proclaimed him an inheritor of the unvitiated blood of those proud warrior hunters, who, in quest of the great New England moose, had scoured, bow in hand, the aboriginal forests of the main. But no longer snuffing in the trail of the wild beasts of the woodland, Tashtego now hunted in the wake of the great whales of the sea; the unerring harpoon of the son fitly replacing the infallible arrow of the sires. Tashtego was Stubb the second mate's squire.

Third among the harpooneers was Daggoo, a gigantic, coal-black negro-savage, with a lion-like tread. Suspended from his ears were two golden hoops, so large that the sailors called them ring-bolts, and would talk of securing the top-sail halyards to them. In his youth Daggoo had voluntarily shipped on board of a whaler, lying in a lonely bay on his native coast. And never having been anywhere in the world but in Africa, Nantucket, and the pagan harbors most frequented by whalemen, Daggoo retained all his barbaric virtues; and erect as a giraffe, moved about the decks in all the pomp of six feet five in his socks. Curious to tell, this imperial negro, Ahasuerus Daggoo, was the squire of little Flask, who looked like a chess-man beside him. As for the residue of the Pequod's company, be it said, that at the present day not one in two of the many thousand men before the mast employed in the American whale fishery, are Americans born, though pretty nearly all the officers are. Herein it is the same with the American whale fishery as with the American army and military and merchant navies, and the engineering forces employed in the construction of the American Canals and Railroads. The same, I say, because in all these cases the native American liberally provides the brains, the rest of the world as generously supplying the muscles. No small number of these whaling seamen belong to the Azores, where the outward bound Nantucket whalers frequently touch to augment their crews from the hardy peasants of those rocky shores. How it is, there is no telling, but Islanders seem to make the best whalemen. They were nearly all Islanders in the Pequod, *Isolatoes* too, I call such, not acknowledging the common continent of men, but each *Isolato* living on a separate continent of his own.

Whaling crew, circa 1920. *In contrast to Melville's "brawny Labrador sea dogs," these are more normal-looking working men. In most American whaling ports, from the mid-19th century to the 1920s, crews were predominately made up of young men between 15 and 25 who were in desperate need of work. Many of the crew were men of color or immigrants; some were illegal aliens or fugitives; and still others, like Herman Melville, were inexperienced, American-born, adventure-seekers.* — Pardon B. Gifford photograph

Kendall Collection, NBWM

Chapter XXVIII

Ahab

Ahab. — *by Rockwell Kent. ©1930 by R. R. Donnelley & Sons Co., with permission.*

For several days after leaving Nantucket, nothing above hatches was seen of Captain Ahab. The mates regularly relieved each other at the watches, and for aught that could be seen to the contrary, they seemed to be the only commanders of the ship; only they sometimes issued from the cabin with orders so sudden and peremptory, that after all it was plain they but commanded vicariously. Yes, their supreme lord and dictator was there, though hitherto unseen by any eyes not permitted to penetrate into the now sacred retreat of the cabin.

Now, it being Christmas when the ship shot from out her harbor, for a space we had biting Polar weather, though all the time running away from it to the southward; and by every degree and minute of latitude which we sailed, gradually leaving that merciless winter, and all its intolerable weather behind us. It was one of those less lowering, but still grey and gloomy enough mornings of the transition, when with a fair wind the ship was rushing through the water with a vindictive sort of leaping and melancholy rapidity, that as I mounted to the deck at the call of the forenoon watch, so soon as I levelled my glance towards the taffrail, foreboding shivers ran over me. Reality outran apprehension; Captain Ahab stood upon his quarter-deck.

There seemed no sign of common bodily illness about him, nor of the recovery from any. He looked like a man cut away from the stake, when the fire has overrunningly wasted all the limbs without consuming them, or taking away one particle from their compacted aged robustness. His whole high, broad form, seemed made of solid bronze, and shaped in an unalterable mould, like Cellini's cast Perseus. Threading its way out from among his grey hairs, and continuing right down one side of his tawny scorched face and neck, till it disappeared in his clothing, you saw a slender rod-like mark, lividly whitish. Whether that mark was born with him, or whether it was the scar left by some desperate wound, no one could certainly say. By some tacit consent, throughout the voyage little or no allusion was made to it, especially by the mates.

So powerfully did the whole grim aspect of Ahab affect me, and the livid brand which streaked it, that for the first few moments I hardly noted that not a little of this overbearing grimness was owing to the barbaric white leg upon which he partly stood. It had previously come to me that this ivory leg had at sea been fashioned from the polished bone of the sperm whale's jaw.

I was struck with the singular posture he maintained. Upon each side of the Pequod's quarter deck, and pretty close to the mizen shrouds, there was an auger hole, bored about half an inch or so, into the plank. His bone leg steadied in that hole; one arm elevated, and holding by a shroud; Captain Ahab stood erect, looking straight out beyond the ship's ever-pitching prow.

Ere long, from his first visit in the air, he withdrew into his cabin. But after that morning, he was every day visible to the crew; either standing in his pivot-hole, or seated upon an ivory stool he had; or heavily walking the deck. And, by and by, it came to pass, that he was almost continually in the air; but, as yet, for all that he said, or perceptibly did, on the at last sunny deck, he seemed as unnecessary there as another mast. But the Pequod was only making a passage now; not regularly cruising; nearly all whaling preparatives needing supervision the mates were fully competent to, so that there was little or nothing, out of himself, to employ or excite Ahab, now; and thus chase away, for that one interval, the clouds that layer upon layer were piled upon his brow, as ever all clouds choose the loftiest peaks to pile themselves upon.

Nevertheless, ere long, the warm, warbling persuasiveness of the pleasant, holiday weather we came to, seemed gradually to charm him from his mood. More than once did he put forth the faint blossom of a look, which, in any other man, would have soon flowered out in a smile.

Chapter XXIX

Enter Ahab; to him, Stubb

A day in the life of officers aboard whaleship *Lucretia*, cruising for whales along the Northwest Coast, 1887. — *Herbert L. Aldrich photograph. New Bedford Free Public Library.*

Some days elapsed, and ice and icebergs all astern, the Pequod now went rolling through the bright Quito spring, which, at sea, almost perpetually reigns on the threshold of the eternal August of the Tropic. The warmly cool, clear, ringing, perfumed, overflowing, redundant days, were as crystal goblets of Persian sherbet, heaped up—flaked up, with rose-water snow. The starred and stately nights seemed haughty dames in jewelled velvets, nursing at home in lonely pride, the memory of their absent conquering Earls, the golden helmeted suns! For sleeping man, 'twas hard to choose between such winsome days and seducing nights. But all the witcheries of that unwaning weather did not merely lend new spells and potencies to the outward world. Inward they turned upon the soul, especially when the still mild hours of eve came on; then, memory shot her crystals as the clear ice most forms of noiseless twilights. And all these subtle agencies, more and more they wrought on Ahab's texture.

Old age is always wakeful; as if, the longer linked with life, the less man has to do with aught that looks like death. Among sea-commanders, the old greybeards will oftenest leave their berths to visit the night-cloaked deck. It was so with Ahab; only that now, of late, he seemed so much to live in the open air, that truly speaking, his visits were more to the cabin, than from the cabin to the planks. "It feels like going down into one's tomb,"—he would mutter to himself,— "for an old captain like me to be descending this narrow scuttle, to go to my grave-dug berth."

So, almost every twenty-four hours, when the watches of the night were set, and the band on deck sentinelled the slumbers of the band below; and ere long the old man would emerge, griping at the iron banister, to help his crippled way. Some considerating touch of humanity was in him; for at times like these, he usually abstained from patrolling the quarter-deck; because to his wearied mates, seeking repose within six inches of his ivory heel, such would have been the reverberating crack and din of that bony step, that their dreams would have been of the crunching teeth of sharks. But once, the mood was on him too deep for common regardings; and as with heavy, lumber-like pace he was measuring the ship from taffrail to mainmast, Stubb, the odd second mate, came up from below, and with a certain unassured, deprecating humorousness, hinted that if Captain Ahab was pleased to walk the planks, then, no one could say nay; but there might be some way of muffling the noise; hinting something indistinctly and hesitatingly about a globe of tow, and the insertion into it, of the ivory heel. Ah! Stubb, thou did'st not know Ahab then.

"Am I a cannon-ball, Stubb," said Ahab, "that thou wouldst wad me that fashion? But go thy ways; I had forgot. Below to thy nightly grave; where such as ye sleep between shrouds, to use ye to the filling one at last.—Down, dog, and kennel!"

Stubb was speechless a moment; then said excitedly, "I am not used to be spoken to that way, sir; I do but less than half like it, sir."

"Avast!" gritted Ahab between his set teeth, and violently moving away, as if to avoid some passionate temptation.

"No, sir; not yet," said Stubb, emboldened, "I will not tamely be called a dog, sir."

"Then be called ten times a donkey, and a mule, and an ass, and begone, or I'll clear the world of thee!"

As he said this, Ahab advanced upon him with such overbearing terrors in his aspect, that Stubb involuntarily retreated.

"I was never served so before without giving a hard blow for it," muttered Stubb, as he found himself descending the cabin-scuttle. "It's very queer. Stop, Stubb; somehow, now, I don't well know whether to go back and strike him, or—what's that?—down here on my knees and pray for him? Yes, that was the thought coming up in me; but it would be the first time I ever *did* pray. It's queer; very queer; and he's queer too; aye, take him fore and aft, he's about the queerest old man Stubb ever sailed with. He's full of riddles; I wonder what he goes into the after hold for, every night, as Dough-Boy tells me he suspects; what's that for, I should like to know? Who's made appointments with him in the hold? Ain't that queer, now? But how's that? didn't he call me a dog? blazes! he called me ten times a donkey, and piled a lot of jackasses on top of that! He might as well have kicked me, and done with it. Maybe he did kick me, and I didn't observe it, I was so taken all aback with his brow, somehow. It flashed like a bleached bone. What the devil's the matter with me? I don't stand right on my legs. Coming afoul of that old man has a sort of turned me wrong side out. By the Lord, I must have been dreaming, though—how? how? how?—but the only way's to stash it; so here goes to hammock again; and in the morning, I'll see how this plaguey juggling thinks over by day-light."

Whaleship Charles W. Morgan, circa 1918. — *Arthur F. Packard photograph*

Courtesy of T. E. Holcombe

Chapter XXX, XXXI

The Pipe · Queen Mab

"The Cook and the Pilot," 1860.
— *Harper's New Monthly Magazine.*

When Stubb had departed, Ahab stood for a while leaning over the bulwarks; and then, as had been usual with him of late, calling a sailor of the watch, he sent him below for his ivory stool, and also his pipe. Lighting the pipe at the binnacle lamp and planting the stool on the weather side of the deck, he sat and smoked.

Some moments passed, during which the thick vapor came from his mouth in quick and constant puffs, which blew back again into his face. "How now," he soliloquized at last, withdrawing the tube, "this smoking no longer soothes. Oh, my pipe! hard must it go with me if thy charm be gone! What business have I with this pipe? This thing that is meant for sereneness, to send up mild white vapors among mild white hairs, not among torn iron-grey locks like mine. I'll smoke no more—"

He tossed the still lighted pipe into the sea. The fire hissed in the waves; the same instant the ship shot by the bubble the sinking pipe made. With slouched hat, Ahab lurchingly paced the planks.

CHAPTER XXXI

Next morning Stubb accosted Flask. "Such a queer dream, King-Post, I never had. You know the old man's ivory leg, well I dreamed he kicked me with it; and when I tried to kick back, upon my soul, my little man, I kicked my leg right off! And then, presto! Ahab seemed a pyramid, and I, like a blazing fool, kept kicking at it. But what was still more curious, Flask, I somehow seemed to be thinking to myself, that after all, it was not much of an insult, that kick from Ahab. "Why," thinks I, "what's the row? It's not a real leg, only a false leg." And there's a mighty difference between a living thump and a dead thump. That's what makes a blow from the hand, Flask, fifty times more savage to bear than a blow from a cane. I was thinking to myself, "what's his leg now, but a cane—a whalebone cane." But now comes the greatest joke of the dream, Flask. While I was battering away at the pyramid, a sort of badger-haired old merman, with a hump on his back, takes me by the shoulders, and slews me round. "Look ye here," says he; "let's argue the insult. Captain Ahab kicked ye, didn't he?" "Yes, he did," says I— "Well then," says he, "wise Stubb, what have you to complain of? Didn't he kick with right good will? it wasn't a common pitch pine leg he kicked with, was it? No, you were kicked by a great man, and with a beautiful ivory leg, Stubb. It's an honor; account his kicks honors; and on no account kick back; for you can't help yourself, wise Stubb. Don't you see that pyramid?" With that, he all of a sudden seemed to swim off into the air. I snored; rolled over; and there I was in my hammock! Now, what do you think of that dream, Flask?"

"I don't know; it seems a sort of foolish to me, tho'."

"May be. But it's made a wise man of me, Flask.

"D'ye see Ahab standing there, sideways looking over the stern? Well, the best thing you can do, Flask, is to let that old man alone; never speak to him, whatever he says. Halloa! what's that he shouts? Hark!"

"Mast-head, there! Look sharp, all of ye! There are whales hereabouts! If ye see a white one, split your lungs for him!"

"What d'ye think of that now, Flask? ain't there a small drop of something queer about that, eh? a white whale—did ye mark that, man? Look ye—there's something special in the wind. Stand by for it, Flask. Ahab has that that's bloody on his mind. But, mum; he comes this way."

Whaleship **Charles W. Morgan,** *circa 1920.*

William R. Hegarty Collection

Chapter XXXII

Cetology

"Three sperm whales in various attitudes." Melville wrote that the whale in the center of Beale's illustration was "wretchedly cropped and dwarfed, and looks altogether unnatural. The head is good." — *Source: Leyda. Engraving from Beale.*

t is some systematized exhibition of the whale in his broad genera, that I would now fain put before you. Yet is it no easy task. The classification of the constituents of a chaos, nothing less is here essayed.

Many are the men, small and great, old and new, landsmen and seamen, who have at large or in little, written of the whale.

There are only two books in being which at all pretend to put the living sperm whale before you, and at the same time, in the remotest degree succeed in the attempt. Those books are Beale's and Bennett's; both in their time surgeons to English South-Sea whale-ships, and both exact and reliable men. The original matter touching the sperm whale to be found in their volumes is necessarily small; but so far as it goes, it is of excellent quality, though mostly confined to scientific description. As yet, however, the sperm whale, scientific or poetic, lives not complete in any literature. Far above all other hunted whales, his is an unwritten life.

Now the various species of whales need some sort of popular comprehensive classification, if only an easy outline one for the present, hereafter to be filled in all its departments by subsequent laborers. But it is a ponderous task; no ordinary letter-sorter in the Post-office is equal to it.

First: The uncertain, unsettled condition of this science of Cetology is in the very vestibule attested by the fact, that in some quarters it still remains a moot point whether a whale be a fish.

Be it known that, waiving all argument, I take the good old fashioned ground that the whale is a fish, and call upon holy Jonah to back me. This fundamental thing settled, the next point is, in what internal respect does the whale differ from other fish. In brief: lungs and warm blood; whereas, all other fish are lungless and cold blooded.

Now, then, come the grand divisions of the entire whale host.

Greenland whale, 1837. *"And here be it said, that the Greenland Whale is an usurper upon the throne of the seas. He is not even by any means the largest of the whales. Reference to nearly all the leviathanic allusions in the great poets of past days, will satisfy you that the Greenland Whale, without one rival, was to them the monarch of the seas. But the time has at last come for a new proclamation. This is Charing Cross; hear ye! good people all,—the Greenland Whale is deposed,—the great Sperm Whale now reigneth!"* —Chapter 32. Engraving by Robert Hamilton: Natural History of…Whales.

New Bedford Whaling Museum

Sperm Whale — This whale is, without doubt, the largest inhabitant of the globe; the most formidable of all whales to encounter; the most majestic in aspect; and lastly, by far the most valuable in commerce; he being the only creature from which that valuable substance, spermaceti, is obtained.

Some centuries ago, when the sperm whale was almost wholly unknown in his own proper individuality, and when his oil was only accidentally obtained from the stranded fish; in those days spermaceti, it would seem, was popularly supposed to be derived from a creature identical with the one then known in England as the Greenland or right whale. It was the idea also, that this same spermaceti was that quickening humor of the Greenland whale which the first syllable of the word literally expresses. In those times, also, spermaceti was exceedingly scarce, not being used for light, but only as an ointment and medicament. It was only to be had from the druggists as you nowadays buy an ounce of rhubarb. When, as I opine, in the course of time, the true nature of spermaceti became known, its original name was still retained by the dealers; no doubt to enhance its value by a notion so strangely significant of its scarcity. And so the appellation must at last have come to be bestowed upon the whale from which this spermaceti was really derived.

Right Whale — In one respect this is the most venerable of the leviathans, being the one first regularly hunted by man. It yields the article commonly known as whalebone or baleen; and the oil specially known as "whale oil," an inferior article in commerce.

Some pretend to see a difference between the Greenland whale of the English and the right whale of the Americans. But they precisely agree in all their grand features; nor has there yet been presented a single determinate fact upon which to ground a radical distinction.

Right whale (top) and humpback whale, 1851. *These two whales were first sketched in 1847 by John Manjiro aboard the ship* Franklin *on the Kubatsu Bay whaling grounds near the Sea of Japan. Manjiro wrote: "The right whale is well-fed, its length is a little longer than the humpback."* — *Watercolor by Shoryo Kawada:* Hyoson Kiryaku.

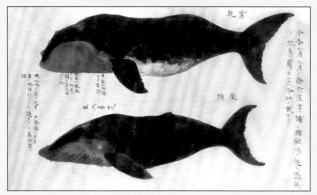

Millicent Library, Fairhaven

The dying humpback, Kodiak, Alaska, circa 1920. — *Spinner Collection.*

Fin-Back — In the length he attains, and in his baleen, the fin-back resembles the right whale, but is of a less portly girth, and a lighter color, approaching to olive. His great lips present a cable-like aspect, formed by the intertwisting, slanting folds of large wrinkles. His grand distinguishing feature, the fin, from which he derives his name, is often a conspicuous object. This fin is some three or four feet long, growing vertically from the hinder part of the back, of an angular shape, and with a very sharp pointed end. The fin-back is not gregarious. He seems a whale-hater, as some men are man-haters. Very shy; always going solitary; unexpectedly rising to the surface in the remotest and most sullen waters; his straight and single lofty jet rising like a tall misanthropic spear upon a barren plain; gifted with such wondrous power and velocity in swimming, as to defy all present pursuit from man.

Hump Back — This whale is often seen on the northern American coast. He has been frequently captured there, and towed into harbor. He has a great pack on him like a peddler; or you might call him the elephant and castle whale. His oil is not very valuable. He has baleen. He is the most gamesome and light-hearted of all the whales, making more gay foam and white water generally than any other of them.

Sulphur Bottom — Another retiring gentleman, with a brimstone belly, doubtless got by scraping along the Tartarian tiles in some of his profounder divings. He is seldom seen. He is never chased; he would run away with rope-walks of line. Prodigies are told of him. Adieu, sulphur bottom! I can say nothing more that is true of ye, nor can the oldest Nantucketer.

Grampus — Though this fish, whose loud sonorous breathing, or rather blowing, has furnished a proverb to landsmen, is so well known a denizen of the deep, yet is he not popularly classed among whales. He is of moderate size, varying from fifteen to twenty-five feet in length, and of corresponding dimensions round the waist. He swims in herds; he is never regularly hunted, though his oil is considerable in quantity, and pretty good for light. By some fishermen his approach is regarded as premonitory of the advance of the great sperm whale.

Black Fish — Call him the hyena whale, if you please. His voracity is well known, and from the circumstance that the inner angles of his lips are curved upwards, he carries an everlasting Mephisthelean grin on his face. This whale averages some sixteen or eighteen feet in length. He is found in almost all latitudes. He has a peculiar way of showing his dorsal hooked fin in swimming, which looks something like a Roman nose. When not more profitably employed, the sperm whale hunters sometimes capture the hyena whale, to keep up the supply of cheap oil for domestic employment—as some frugal housekeepers, in the absence of company, and quite alone by themselves, burn unsavory tallow instead of odorous wax. Though their blubber is very thin, some of these whales will yield you upwards of thirty gallons of oil.

Narwhale, that is, nostril whale — Another instance of a curiously named whale, so named I suppose from his peculiar horn being originally mistaken for a peaked nose. The creature is some sixteen feet in length, while its horn averages five feet, though some exceed ten, and even attain to fifteen feet. Strictly speaking, this horn is but a lengthened tusk. What precise purpose this ivory horn or lance answers, it would be hard to say. Charley Coffin said it was used for an ice-piercer; for the narwhale, rising to the surface of the Polar Sea, and finding it sheeted with ice, thrusts his horn up, and so breaks through. From certain cloistered old authors I have

Blackfish ashore at Nantucket, 1870. *These small whales congregate along coastal waters to feed on fish and squid. When blackfish were spotted near shore, men in small boats would get between them and deep water. By making noise, the men would "gallie," or frighten, the whales into beaching. The whales were then sold to an oil works company, such as William F. Nye in New Bedford.* — Joseah Freeman photograph.

New Bedford Whaling Museum

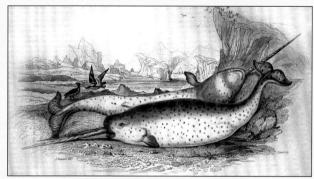

Narwhale or unicorn whale, 1837. — Engraving by Robert Hamilton: Natural History of…Whales. New Bedford Whaling Museum.

gathered that this same sea-unicorn's horn was in ancient days regarded as the great antidote against poison, and as such, preparations of it brought immense prices. It was also distilled to a volatile salts for fainting ladies, the same way that the horns of the male deer are manufactured into hartshorn.

The narwhale has a very picturesque, leopard-like look, being of a milk-white ground color, dotted with round and oblong spots of black. His oil is very superior, clear and fine; but there is little of it, and he is seldom hunted. He is mostly found in the circumpolar seas.

Killer — Of this whale little is precisely known to the Nantucketer, and nothing at all to the professed naturalist. From what I have seen of him at a distance, I should say that he was about the bigness of a grampus. He is very savage—a sort of Feegee fish. He sometimes takes the great Folio Whales by the lip, and hangs there like a leech, till the mighty brute is worried to death. The Killer is never hunted. I never heard what sort of oil he has. Exception might be taken to the name bestowed upon this whale, on the ground of its indistinctness. For we are all killers, on land and on sea; Bonapartes and Sharks included.

Thrasher — This gentleman is famous for his tail, which he uses for a ferule in thrashing his foes. He mounts the folio whale's back, and as he swims, he works his passage by flogging him; as some schoolmasters get along in the world by a similar process. Still less is known of the thrasher than of the killer. Both are outlaws, even in the lawless seas.

I now leave my cetological System standing thus unfinished, even as the great Cathedral of Cologne was left, with the crane still standing upon the top of the uncompleted tower. For small erections may be finished by their first architects; grand ones, true ones, ever leave the copestone to posterity. God keep me from ever completing anything. This whole book is but a draught—nay, but the draught of a draught. Oh Time, Strength, Cash, and Patience!

Chapter XXXIII, XXXIV

The Specksynder · The Cabin Table

"Life in the Forecastle," 1846. — *From Davis:* Nimrod of the Sea.

Concerning the officers of the whale-craft, this seems as good a place as any to set down a little domestic peculiarity on ship-board, arising from the existence of the harpooneer class of officers, a class unknown of course in any other marine than the whale-fleet.

The large importance attached to the harpooneer's vocation is evinced by the fact, that originally in the old Dutch Fishery, two centuries and more ago, the command of a whale ship was not wholly lodged in the person now called the captain, but was divided between him and an officer called the Specksynder. Literally this word means Fat-Cutter; usage, however, in time made it equivalent to Chief Harpooneer. In those days, the captain's authority was restricted to the navigation and general management of the vessel: while over the whale-hunting department and all its concerns, the Specksynder or Chief Harpooneer reigned supreme. At present he ranks simply as senior Harpooneer; and as such, is but one of the captain's more inferior subalterns. Nevertheless, as upon the good conduct of the harpooneers the success of a whaling voyage largely depends, and since in the American Fishery he is not only an important officer in the boat, but under certain circumstances (night watches on a whaling ground) the command of the ship's deck is also his; therefore the grand political maxim of the sea demands, that he should nominally live apart from the men before the mast, and be in some way distinguished as their professional superior; though always, by them, familiarly regarded as their social equal.

Now, the grand distinction drawn between officer and man at sea, is this—the first lives aft, the last forward. Hence, in whale-ships and merchantmen alike, the mates have their quarters with the captain; and so, too, in most of the American whalers the harpooneers are lodged in the after part of the ship. That is to say, they take their meals in the captain's cabin, and sleep in a place indirectly communicating with it.

CHAPTER XXXIV

It is not the least among the strange things bred by the intense artificialness of sea-usages, that while in the open air of the deck some officers will, upon provocation, bear themselves boldly and defyingly enough towards their commander; yet, ten to one, let those very officers the next moment go down to their customary dinner in that same commander's cabin, and straightway their inoffensive, not to say deprecatory and humble air towards him, as he sits at the head of the table; this is marvellous, sometimes most comical. Wherefore this difference? Who has but once dined his friends, has tasted what it is to be Caesar. It is a witchery of social czarship which there is no withstanding. Now, if to this consideration you superadd the official supremacy of a ship-master, then, by inference, you will derive the cause of that peculiarity of sea-life just mentioned.

Crew of the whaleship Baltic, 1859. *This sketch is one in a series of ten that appeared as engravings in an accompanying story in* Harper's Magazine. *These sailors just arrived in New Bedford and are being preyed upon by "land sharks," (in background), a "set of small traders, agents… pimps, etc., who pander to the vices of the outgoing or returning seamen."* — Sketch by David Hunter Strother.

Kendall Collection, NBWM

Now, Ahab and his three mates formed what may be called the first table in the Pequod's cabin. After their departure, taking place in inverted order to their arrival, the canvas cloth was cleared, or rather was restored to some hurried order by the pallid steward. And then the three harpooneers were bidden to the feast, they being its residuary legatees. They made a sort of temporary servants' hall of the high and mighty cabin.

In strange contrast to the hardly tolerable constraint and nameless invisible domineerings of the captain's table, was the entire care-free license and ease, the almost frantic democracy of those inferior fellows the harpooneers. While their masters, the mates, seemed afraid of the sound of the hinges of their own jaws, the harpooneers chewed their food with such a relish that there was a report to it. They dined like lords; they filled their bellies like Indian ships all day loading with spices.

But, though these barbarians dined in the cabin, and nominally lived there; still, being anything but sedentary in their habits, they were scarcely ever in it except at meal-times, and just before sleeping-time, when they passed through it to their own peculiar quarters.

In this one matter, Ahab seemed no exception to most American whale captains, who, as a set, rather incline to the opinion that by rights the ship's cabin belongs to them; and that it is by courtesy alone that anybody else is, at any time, permitted there. So that, in real truth, the mates and harpooneers of the Pequod might more properly be said to have lived out of the cabin than in it. Nor did they lose much hereby; in the cabin was no companionship; socially, Ahab was inaccessible. Though nominally included in the census of Christendom, he was still an alien to it.

Forecastle of the* Charles W. Morgan, *1922. *On an American whaler, 15 to 20 men shared crowded living quarters in the forecastle, which was dirty, smelly, slimy and wet. Their food was stale or rancid and the air foul and hot. The salty group pictured here are American actors. Real-life crews often hailed from a variety of countries and spoke different languages and dialects.* — From the motion picture, "Down to the Sea in Ships."

New Bedford Whaling Museum

Chapter XXXV

The Lookout

It was during the more pleasant weather, that in due rotation with the other seamen my first mast-head came round.

In most American whalemen the mast-heads are manned almost simultaneously with the vessel's leaving her port; even though she may have fifteen thousand miles, and more, to sail ere reaching her proper cruising ground. And if, after a three, four, or five years' voyage she is drawing nigh home with anything empty in her—say, an empty vial even—then, her mast-heads are kept manned to the last; and not till her skysail-poles sail in among the spires of the port, does she altogether relinquish the hope of capturing one whale more.

In search of whales, circa 1920.

Kendall Collection, NBWM

Crow's Nest, *Charles W. Morgan*.
— From "Down to the Sea in Ships." New Bedford Whaling Museum.

The three mast-heads are kept manned from sun-rise to sun-set; the seamen taking their regular turns (as at the helm), and relieving each other every two hours. In the serene weather of the tropics it is exceedingly pleasant the mast-head; nay, to a dreamy meditative man it is delightful. There you stand, a hundred feet above the silent decks, striding along the deep, as if the masts were gigantic stilts, while beneath you and between your legs, as it were, swim the hugest monsters of the sea, even as ships once sailed between the boots of the famous Colossus at old Rhodes. There you stand, lost in the infinite series of the sea, with nothing ruffled but the waves. The tranced ship indolently rolls; the drowsy trade winds blow; everything resolves you into languor. For the most part, in this tropic whaling life, a sublime uneventfulness invests you; you hear no news; read no gazettes; extras with startling accounts of commonplaces never delude you into unnecessary excitements; you hear of no domestic afflictions; bankrupt securities; fall of stocks; are never troubled with the thought of what you shall have for dinner—for all your meals for three years and more are snugly stowed in casks, and your bill of fare is immutable.

Your most usual point of perch is the head of the t' gallant-mast, where you stand upon two thin parallel sticks (almost peculiar to whalemen) called the t' gallant cross-trees. Here, tossed about by the sea, the beginner feels about as cosy as he would standing on a bull's horns. To be sure, in cold weather you may carry your house aloft with you, in the shape of a watch-coat; but properly speaking the thickest watch-coat is no more of a house than the unclad body; so a watch-coat is not so much of a house as it is a mere envelope, or additional skin encasing you.

For one, I used to lounge up the rigging very leisurely, resting in the top to have a chat with Queequeg, or any one else off duty whom I might find there; then ascending a little way further, and throwing a lazy leg over the top-sail yard, take a preliminary view of the watery pastures, and so at last mount to my ultimate destination.

Chapter XXXVI

The Quarter-Deck

It was not a great while after the affair of the pipe, that one morning shortly after breakfast, Ahab, as was his wont, ascended the cabin-gangway to the deck. There most sea-captains usually walk at that hour, as country gentlemen, after the same meal, take a few turns in the garden.

Soon his steady, ivory stride was heard, as to and fro he paced his old rounds, upon planks so familiar to his tread, that they were all over dented, like geological stones, with the peculiar mark of his walk. Did you fixedly gaze, too, upon that ribbed and dented brow; there also, you would see still stranger foot-prints—the foot-prints of his one unsleeping, ever-pacing thought.

But on the occasion in question, those dents looked deeper, even as his nervous step that morning left a deeper mark. And, so full of his thought was Ahab, that at every uniform turn that he made, now at the main-mast and now at the binnacle, you could almost see that thought turn in him as he turned, and pace in him as he paced; so completely possessing him, indeed, that it all but seemed the inward mould of every outer movement.

"D'ye mark him, Flask?" whispered Stubb; "the chick that's in him pecks the shell. T'will soon be out."

The hours wore on;—Ahab now shut up within his cabin; anon, pacing the deck, with the same intense bigotry of purpose in his aspect.

It drew near the close of day. Suddenly he came to a halt by the bulwarks, and inserting his bone leg into the augerhole there, and with one hand grasping a shroud, he ordered Starbuck to send everybody aft.

"Sir!" said the mate, astonished at an order seldom or never given on ship-board except in some extraordinary case.

When the entire ship's company were assembled, and with curious and not wholly unapprehensive faces, were eyeing him, for he looked not unlike the weather horizon when a storm is coming up, Ahab, after rapidly glancing over the bulwarks, and then darting his eyes among the crew, started from his standpoint; and as though not a soul were nigh him resumed his heavy turns upon the deck. With bent head and half-slouched hat he continued to pace, unmindful of the wondering whispering among the men; till Stubb cautiously whispered to Flask, that Ahab must have summoned them there for the purpose of witnessing a pedestrian feat. But this did not last long. Vehemently pausing, he cried:—"What do ye do when ye see a whale, men?"

Comrades in blubber, circa 1865. — *Watercolor from anonymous sketchbook aboard bark* Orray Taft *of New Bedford. Kendall Collection, NBWM.*

"Sing out for him!" was the impulsive rejoinder from a score of clubbed voices. "Good!" cried Ahab, with a wild approval in his tones; observing the hearty animation into which his unexpected question had so magnetically thrown them.

"And what do ye next, men?"

"Lower away, and after him!"

"And what tune is it ye pull to, men?"

"A dead whale or a stove boat!"

More and more strangely and fiercely glad and approving, grew the countenance of the old man at every shout; while the mariners began to gaze curiously at each other, as if marvelling how it was that they themselves became so excited at such seemingly purposeless questions.

But, they were all eagerness again, as Ahab, now half-revolving in his pivot-hole, with one hand reaching high up a shroud, and tightly, almost convulsively grasping it, addressed them thus:—

"All ye mast-headers have before now heard me give orders about a white whale. Look ye! d'ye see this Spanish ounce of gold?—holding up a broad bright coin to the sun—it is a sixteen dollar piece, men. D'ye see it? Mr. Starbuck, hand me yon top-maul."

Receiving the top-maul from Starbuck, he advanced towards the main-mast with the hammer uplifted in one hand, exhibiting the gold with the other, and with a high raised voice exclaiming: "Whosoever of ye raises me a white-headed whale with a wrinkled brow and a crooked jaw; whosoever of ye raises me that white-headed whale, with three holes punctured in his starboard fluke—look ye, whosoever of ye raises me that same white whale, he shall have this gold ounce, my boys!"

"Huzza! huzza!" cried the seamen, as with swinging tarpaulins they hailed the act of nailing the gold to the mast.

"Captain Ahab," said Tashtego, "that white whale must be the same that some call Moby Dick."

"Moby Dick?" shouted Ahab. "Do ye know the white whale then, Tash?"

"Captain Ahab," said Starbuck, who, with Stubb and Flask, had thus far been eyeing his superior with increasing surprise, but at last seemed struck with a thought which somewhat explained all the wonder. "Captain Ahab, I have heard of Moby Dick—but it was not Moby Dick that took off thy leg?"

"Who told thee that?" cried Ahab; then pausing, "Aye, Starbuck; aye, my hearties all round; it was Moby Dick that dismasted me; Moby Dick that brought me to this dead stump I stand on now. Aye, aye," he shouted with a terrific, loud, animal sob, like that of a heart-stricken moose; "Aye, aye! it was that accursed white whale that razeed me; made a poor pegging lubber of me for ever and a day!" Then tossing both arms, with measureless imprecations he shouted out: "Aye, aye! and I'll chase him round Good Hope, and round the Horn, and round the Norway maelstrom, and round perdition's flames before I give him up. And this is what ye have shipped for, men! to chase that white whale on both sides of land, and over all sides of earth, till he spouts black blood and rolls fin out. What say ye, men, will ye splice hands on it, now? I think ye do look brave."

"Aye, aye!" shouted the harpooneers and seamen, running closer to the excited old man: "A sharp eye for the White Whale; a sharp lance for Moby Dick!"

"God bless ye," he seemed to half sob and half shout. "God bless ye, men. Steward! go draw the great measure of grog. But what's this long face about, Mr. Starbuck; wilt thou not chase the white whale? Art not game for Moby Dick?"

"I am game for his crooked jaw, and for the jaws of Death too, Captain Ahab, if it fairly comes in the way of the business we follow; but I came here to hunt whales, not my commander's vengeance. How many barrels will thy vengeance yield thee even if thou gettest it, Captain Ahab? It will not fetch thee much in our Nantucket market."

"Nantucket market! Hoot! But come closer, Starbuck; thou requirest a little lower layer. If money's to be the measurer, man, and the accountants have computed their great counting-house the globe, by girdling it with guineas, one to every three parts of an inch; then, let me tell thee, that my vengeance will fetch a great premium *here*!"

"He smites his chest," whispered Stubb, "what's that for? methinks it rings most vast, but hollow."

"Vengeance on a dumb brute!" cried Starbuck, "that simply smote thee from blindest instinct! Madness! To be enraged with a dumb thing, Captain Ahab, seems blasphemous."

"Hark ye yet again,—the little lower layer. All visible objects, man, are but as pasteboard masks. But in each event—in the living act, the undoubted deed—there, some unknown but still reasoning thing puts forth the mouldings of its features from behind the unreason-

Shaping up the last crew of the Charles W. Morgan, 1920. *Captain John Gonsalves (in suit), a Cape Verdean-American, was the last master of the Morgan. The ship had trouble putting together a crew in New Bedford and finally shipped out of Provincetown in September 1920. She later discharged several of the crew in the Dominican Republic. In May of 1921, the ship returned with 2,702 barrels of oil.* — Arthur F. Packard photograph.

Courtesy of T. E. Holcombe

ing mask. If man will strike, strike through the mask! How can the prisoner reach outside except by thrusting through the wall? To me, the white whale is that wall, shoved near to me. Sometimes I think there's naught beyond. But 'tis enough. He tasks me; he heaps me; I see in him outrageous strength, with an inscrutable malice sinewing it. That inscrutable thing is chiefly what I hate; and be the white whale agent, or be the white whale principal, I will wreak that hate upon him. Talk not to me of blasphemy, man; I'd strike the sun if it insulted me."

"God keep me!—keep us all!" murmured Starbuck, lowly.

But in his joy at the enchanted, tacit acquiescence of the mate, Ahab did not hear his foreboding invocation; nor yet the low laugh from the hold; nor yet the presaging vibrations of the winds in the cordage; nor yet the hollow flap of the sails against the masts, as for a moment their hearts sank in. For again Starbuck's downcast eyes lighted up with the stubbornness of life; the subterranean laugh died away; the winds blew on; the sails filled out; the ship heaved and rolled as before.

"The measure! The measure!" cried Ahab.

Receiving the brimming pewter, and turning to the harpooneers, he ordered them to produce their weapons. Then ranging them before him near the capstan, with their harpoons in their hands, while his three mates stood at his side with their lances, and the rest of the ship's company formed a circle round the group; he stood for an instant searchingly eyeing every man of his crew. But those wild eyes met his, as the bloodshot eyes of the prairie wolves meet the eye of their leader, ere he rushes on at their head in the trail of the bison; but, alas! only to fall into the hidden snare of the Indian.

"Drink and pass!" he cried, handing the heavy charged flagon to the nearest seaman. "The crew alone now drink. Round with it, round! Short draughts—long swallows, men; 'tis hot as Satan's hoof. So, so; it goes round excellently. It spiralizes in ye; forks out at the serpent-snapping eye. Well done; almost drained. That way it went, this way it comes. Hand it me—here's a hollow; Men, ye seem the years; so brimming life is gulped and gone. Steward, refill!

"Attend now, my braves. I have mustered ye all round this capstan; and ye mates, flank me with your lances; and ye harpooneers, stand there with your irons; and ye, stout mariners, ring me in, that I may in some sort revive a noble custom of my fisherman fathers before me.

"Advance, ye mates! Cross your lances full before me. Well done! Let me touch the axis." So saying, with extended arm, he grasped the three level, radiating lances at their crossed centre; while so doing, suddenly and nervously twitched them; meanwhile, glancing intently

"Crew on deck," 1846. — *From Davis:* Nimrod of the Sea.

from Starbuck to Stubb; from Stubb to Flask. It seemed as though, by some nameless, interior volition, he would fain have shocked into them the same fiery emotion accumulated within the Leyden jar of his own magnetic life. The three mates quailed before his strong, sustained, and mystic aspect.

"In vain!" cried Ahab; "but, maybe, 'tis well. For did ye three but once take the full-forced shock, then mine own electric thing, *that* had perhaps expired from out me. Perchance, too, it would have dropped ye dead. Perchance ye need it not. Down lances! And now, ye mates, I do appoint ye three cup-bearers to my three pagan kinsmen there. Cut your seizings and draw the poles, ye harpooneers!"

Silently obeying the order, the three harpooneers now stood with the detached iron part of their harpoons, some three feet long, held, barbs up, before him.

"Stab me not with that keen steel! Cant them; cant them over! know ye not the goblet end? Turn up the socket! So, so; now, ye cup-bearers, advance. The irons! take them; hold them while I fill!" Forthwith, slowly going from one officer to the other, he brimmed the harpoon sockets with the fiery waters from the pewter.

"Now, three to three, ye stand. Commend the murderous chalices! Bestow them, ye who are now made parties to this indissoluble league. Ha! Starbuck! but the deed is done! Yon ratifying sun now waits to sit upon it. Drink, ye harpooneers! drink and swear, ye men that man the deathful whaleboat's bow—Death to Moby Dick! God hunt us all, if we do not hunt Moby Dick to his death!" The long, barbed steel goblets were lifted; and to cries and maledictions against the white whale, the spirits were simultaneously quaffed down with a hiss. Starbuck paled, and turned, and shivered. Once more, and finally, the replenished pewter went the rounds among the frantic crew; when, waving his free hand to them, they all dispersed; and Ahab retired within his cabin.

Chapter XXXVII

Sunset

Ship *Huron* of Sag Harbor, 1849. — *From the Gold Rush journal of Amos Chase, Jr., aboard the ship* Constantine.

*The cabin; by the stern windows;
Ahab sitting alone, and gazing out.*

I leave a white and turbid wake; pale waters, paler cheeks, where'er I sail. The envious billows sidelong swell to whelm my track; let them; but first I pass.

Dry heat upon my brow? Oh! time was, when as the sunrise nobly spurred me, so the sunset soothed. No more. This lovely light, it lights not me; all loveliness is anguish to me, since I can ne'er enjoy. Gifted with the high perception, I lack the low, enjoying power; damned, most subtly and most malignantly! damned in the midst of Paradise! Good night—good night! (*Waving his hand, he moves from the window.*)

'Twas not so hard a task. I thought to find one stubborn, at the least; but my one cogged circle fits into all their various wheels, and they revolve. Or, if you will, like so many ant-hills of powder, they all stand before me; and I their match. Oh, hard! that to fire others, the match itself must needs be wasting! What I've dared, I've willed; and what I've willed, I'll do! They think me mad—Starbuck does; but I'm demoniac, I am madness maddened! That wild madness that's only calm to comprehend itself! The prophecy was that I should be dismembered; and—Aye! I lost this leg. I now prophesy that I will dismember my dismemberer. Now, then, be the prophet and the fulfiller one. That's more than ye, ye great gods, ever were. I laugh and hoot at ye, ye cricket-players, ye pugilists, ye deaf Burkes and blinded Bendigoes! I will not say as school-boys do to bullies,—Take some one of your own size; don't pommel *me*! No, ye've knocked me down, and I am up again; but *ye* have run and hidden. Come forth from behind your cotton bags! The path to my fixed purpose is laid with iron rails, whereon my soul is grooved to run. Over unsounded gorges, through the rifled hearts of mountains, under torrents' beds, unerringly I rush! Naught's an obstacle, naught's an angle to the iron way!

Whaling ship William makes wake as it departs the Isle of Mas Afuera near Chile, 1797. "Yonder, by the ever-brimming goblet's rim, the warm waves blush like wine. The gold brow plumbs the blue. The diver sun—slow dived from noon,—goes down; my soul mounts up! she wearies with her endless hill. Is, then, the crown too heavy that I wear? this Iron Crown of Lombardy.... My brain seems to beat against the solid metal; aye, steel skull, mine; the sort that needs no helmet in the most brain-battering fight!" — *Ahab: Chapter 37. Watercolor by Thomas Wetling, from the journal of the ship* William *of London.*

Kendall Collection, NBWM

Chapter XLI

Moby Dick

I, Ishmael, was one of that crew; my shouts had gone up with the rest; my oath had been welded with theirs; and stronger I shouted, and more did I hammer and clinch my oath, because of the dread in my soul. A wild, mystical, sympathetical feeling was in me; Ahab's quenchless feud seemed mine. With greedy ears I learned the history of that murderous monster against whom I and all the others had taken our oaths of violence and revenge.

For some time past, though at intervals only, the unaccompanied, secluded White Whale had haunted those uncivilized seas mostly frequented by the Sperm Whale fishermen. But not all of them knew of his existence; only a few of them, comparatively, had knowingly seen him; while the number who as yet had actually and knowingly given battle to him, was small indeed.

And as for those who, previously hearing of the White Whale, by chance caught sight of him; in the beginning of the thing they had every one of them, almost, as boldly and fearlessly lowered for him, as for any other whale of that species. But at length, such clamities did ensue in thses assaults—not restricted to sprained wrists and ankles, broken limbs, or devouring amputations—but fatal to the last degree of fatality; those repeated disastrous repulses, all accumulating and piling their terrors upon Moby Dick; those things had gone far to shake the fortitude of many brave hunters, to whom the story of the White Whale had eventually come.

Nor did wild rumors of all sorts fail to exaggerate, and still the more horrify the true histories of these deadly encounters. For not only do fabulous rumors naturally grow out of the very body of all surprising terrible events,—as the smitten tree gives birth to its fungi; but, in maritime life, far more than in that of terra firma, wild rumors abound, wherever there is any adequate reality for them to cling to. And as the sea surpasses the land in this matter, so the whale fishery surpasses every other sort of maritime life, in the wonderfulness and fearfulness of the rumors which sometimes circulate there. For not only are whalemen as a body unexempt from that ignorance and superstitiousness hereditary to all sailors; but of all sailors, they are by all odds the most directly brought into contact with whatever is appallingly astonishing in the sea; face to face they not only eye its greatest marvels, but, hand to jaw, give battle to them.

Nevertheless, some there were, who even in the face of these things were ready to give chase to Moby Dick;

Moby Dick rises. — *by Rockwell Kent.*
©1930 by R. R. Donnelley & Sons Co., with permission.

and a still greater number who, chancing only to hear of him distantly and vaguely, without the specific details of any certain calamity, and without superstitious accompaniments, were sufficiently hardy not to flee from the battle if offered.

One of the wild suggestings referred to, as at last coming to be linked with the White Whale in the minds of the superstitiously inclined, was the unearthly conceit that Moby Dick was ubiquitous; that he had actually been encountered in opposite latitudes at one and the same instant of time.

Nor was this conceit altogether without some faint show of superstitious probability. For as the secrets of the currents in the seas have never yet been divulged, even to the most erudite research; so the hidden ways of the Sperm Whale when beneath the surface remain, in great part, unaccountable to his pursuers; and from time to time have originated the most curious and contradictory speculations regarding them, especially concerning the mystic modes whereby, after sounding to a great depth, he transports himself with such vast swiftness to the most widely distant points.

It is a thing well known to both American and English whale-ships, that some whales have been captured far north in the Pacific, in whose bodies have been found the barbs of harpoons darted in the Greenland seas. Nor is it to be gainsaid, that in some of these instances it has been declared that the interval of time between the two assaults could not have exceeded very many days. Hence, by inference, it has been believed by some whalemen, that the nor'west passage, so long a problem to man, was never a problem to the whale.

So that here, in the real living experience of living men, the prodigies related in old times of the inland Strello mountain in Portugal (near whose top there was said to be a lake in which the wrecks of ships floated up to the surface); and that still more wonderful story of the Arethusa fountain near Syracuse (whose waters were believed to have come from the Holy Land by an underground passage); these fabulous narrations are almost fully equalled by the realities of the whaleman.

But even stripped of these supernatural surmisings, there was enough in the earthly make and incontestable character of the monster to strike the imagination with unwonted power. For, it was not so much his uncommon bulk that so much distinguished him from other sperm whales, but, as was elsewhere thrown out—a peculiar snow-white wrinkled forehead, and a high, pyramidical white hump. These were his prominent features; the tokens whereby, even in the limitless, uncharted seas, he revealed his identity, at a long distance, to those who knew him.

The rest of his body was so streaked, and spotted, and marbled with the same shrouded hue, that, in the end, he had gained his distinctive appellation of the white whale; a name, indeed, literally justified by his vivid aspect, when seen gliding at high noon through a dark blue sea, leaving a milky-way wake of creamy foam, all spangled with golden gleamings.

Judge, then, to what pitches of inflamed, distracted fury the minds of his more desperate hunters were impelled, when amid the chips of chewed boats, and the sinking limbs of torn comrades, they swam out of the white curds of the whale's direful wrath into the serene, exasperating sunlight, that smiled on, as if at a birth or a bridal.

His three boats stove around him, and oars and men both whirling in the eddies; one captain, seizing the line-knife from his broken prow, had dashed at the whale, as an Arkansas duellist at his foe, blindly seeking with a six inch blade to reach the fathom-deep life of the whale. That captain was Ahab. And then it was, that suddenly

"*Destruction of the Larboard boat of the* Ann Alexander, *by a sperm whale in the South Pacific.*" *In August 1851, shortly before the release of* Moby-Dick, *the whaleship* Ann Alexander *of New Bedford was whaling in the waters where the* Essex *met her fate nearly two decades earlier. The mate's boat, fast to a whale, suddenly found itself being rushed by its prey. The whale quickly seized the boat in its jaws and smashed it to pieces. The crew was picked up and pursuit continued. Undaunted, the whale again turned upon the mate's boat and crushed it. The captain rescued the swimming crew, returned to the ship and resumed the assault from the foredeck. The captain was able to strike the whale in the head, infuriating it further. After a brief retreat, the whale rushed the ship and struck a blow to her hull; the water came gushing in. Forced to abandon the sinking ship with scant provisions, the crew set upon the open sea. Luckily, they were rescued by the* Nantucket *just two days later. Five months later the* Rebecca Simms *of New Bedford took a whale with two of the* Ann Alexander's *harpoons and pieces of the ship's timbers imbedded in him. Sick and crippled, the whale yielded only 80 barrels of oil. When Melville heard the news, he exclaimed in a letter: "Ye Gods, what a commentator is this* Ann Alexander *whale.... I wonder if my evil art has raised this monster."* — Sources: Ellis, Leyda. Wood engraving by Smyth: Illustrated London News, *11/29/1851.*

Kendall Collection, NBWM

sweeping his sickle-shaped lower jaw beneath him, Moby Dick had reaped away Ahab's leg, as a mower a blade of grass in the field. No turbaned Turk, no hired Venetian or Malay, could have smote him with more seeming malice. Small reason was there to doubt, then, that ever since that almost fatal encounter, Ahab had cherished a wild vindictiveness against the whale. The white whale swam before him as the monomaniac incarnation of all those malicious agencies which some deep men feel eating in them, till they are left living on with half a heart and half a lung. That intangible malignity which has been from the beginning; to whose dominion even the modern Christians ascribe one-half of the worlds;—Ahab did not fall down and worship it like them; but deliriously transferring its idea to the abhorred white whale, he pitted himself, all mutilated, against it. All that most maddens and torments; all that stirs up the lees of things; all truth with malice in it; all that cracks the sinews and cakes the brain; all the subtle demonisms of life and thought; all evil, to crazy Ahab, were visibly personified, and made practically assailable in Moby Dick. He piled upon the whale's white hump the sum of all the general rage and hate felt by his whole race from Adam down; and then, as if his chest had been a mortar, he burst his hot heart's shell upon it.

It is not probable that this monomania in him took its instant rise at the precise time of his bodily dismemberment. Then, in darting at the monster, knife in hand, he had but given loose to a sudden, passionate, corporal animosity; and when he received the stroke that tore him, he probably but felt the agonizing bodily laceration, but nothing more. Yet, when by this collision forced to turn towards home, and for long months of days and weeks, Ahab and anguish lay stretched together in one hammock, rounding in mid winter that dreary, howling Patagonian Cape; then it was, that his torn body and gashed soul bled into one another; and so interfusing, made him mad. That it was only then, on the homeward voyage, after the encounter, that the final monomania seized him, seems all but certain from the fact that, at intervals during the passage, he was a raving lunatic; and, though unlimbed of a leg, yet such vital strength yet lurked in his Egyptian chest, and was moreover intensified by his delirium, that his mates were forced to lace him fast, even there, as he sailed, raving in his hammock. In a strait-jacket, he swung to the mad rockings of the gales. And, when running into more sufferable latitudes, the ship, with mild stun'sails spread, floated across the tranquil tropics, and, to all appearances, the old man's delirium seemed left behind him with the Cape Horn swells, and he came forth from his dark den into the blessed light and air; even then, when he bore that firm, collected front, however pale, and issued his

"Perils Afloat," 1846. "And then it was, that suddenly sweeping his sickle-shaped lower jaw beneath him, Moby Dick had reaped away Ahab's leg, as a mower a blade of grass in the field." — *Chapter 41. Engraving by Alonzo Hartwell. From Delano:* Wanderings and Adventures….

calm orders once again; and his mates thanked God the direful madness was now gone; even then, Ahab, in his hidden self, raved on. Human madness is oftentimes a cunning and most feline thing. When you think it fled, it may have but become transfigured into some still subtler form. Ahab's full lunacy subsided not, but deepeningly contracted; like the unabated Hudson, when that noble Northman flows narrowly, but unfathomably through the Highland gorge. But, as in his narrow-flowing monomania, not one jot of Ahab's broad madness had been left behind; so in that broad madness, not one jot of his great natural intellect had perished. That before living agent, now became the living instrument. If such a furious trope may stand, his special lunacy stormed his general sanity, and carried it, and turned all its concentred cannon upon its own mad mark; so that far from having lost his strength, Ahab, to that one end, did now possess a thousand fold more potency than ever he had sanely brought to bear upon any one reasonable object.

Here, then, was this grey-headed, ungodly old man, chasing with curses a Job's whale round the world, at the head of a crew, too, chiefly made up of mongrel renegades, and castaways, and cannibals—morally enfeebled also, by the incompetence of mere unaided virtue or right-mindedness in Starbuck, the invulnerable jollity of indifference and recklessness in Stubb, and the pervading mediocrity in Flask. Such a crew, so officered, seemed specially picked and packed by some infernal fatality to help him to his monomaniac revenge. How it was that they so aboundingly responded to the old man's ire—by what evil magic their souls were possessed, that at times his hate seemed almost theirs; the white whale as much their insufferable foe as his; what the white whale was to them, or how to their unconscious understandings he might have seemed the gliding great demon of the seas of life,—all this to explain, would be to dive deeper than Ishmael can go.

Chapter XLII

The Whiteness of the Whale

"The Whale's Jaw," 1874. — *From Davis,* Nimrod of the Sea.

What the White Whale was to Ahab, has been hinted; what, at times, he was to me, as yet remains unsaid.

Aside from those more obvious considerations touching Moby Dick, which could not but occasionally awaken in any man's soul some alarm, there was another thought, or rather vague, nameless horror concerning him, which at times by its intensity completely overpowered all the rest; and yet so mystical and well nigh ineffable was it, that I almost despair of putting it in a comprehensible form. It was the whiteness of the whale that above all things appalled me. But how can I hope to explain myself here; and yet, in some dim, random way, explain myself I must, else all these chapters might be naught.

Though in many natural objects, whiteness refiningly enhances beauty, as if imparting some special virtue of its own, as in marbles, japonicas, and pearls; and though various nations have in some way recognised a certain royal pre-eminence in this hue; and though, besides all this, whiteness has been even made significant of gladness, for among the Romans a white stone marked a joyful day; and though in other mortal sympathies and symbolizings, this same hue is made the emblem of many touching, noble things—the innocence of brides, the benignity of age; though among the Red Men of America the giving of the white belt of wampum was the deepest pledge of honor; though in many climes, whiteness typifies the majesty of Justice in the ermine of the Judge, and contributes to the daily state of kings and queens drawn by milk-white steeds; though even in the higher mysteries of the most august religions it has been made the symbol of the divine spotlessness and power; and though directly from the Latin word for white, all Christian priests derive the name of one part of their sacred vesture, the alb or tunic, worn beneath the cassock; yet for all these accumulated associations, with whatever is sweet, and honorable, and sublime, there yet lurks an elusive something in the innermost idea of this hue, which strikes more of panic to the soul than that redness which affrights in blood.

This elusive quality it is, which causes the thought of whiteness, when divorced from more kindly associations, and coupled with any object terrible in itself, to heighten that terror to the furthest bounds. Witness the white bear of the poles, and the white shark of the tropics; what but their smooth, flaky whiteness makes them the transcendent horrors they are? That ghastly whiteness it is which imparts such an abhorrent mildness, even more loathsome than terrific, to the dumb gloating of their aspect. So that not the fierce-fanged tiger in his heraldic coat can so stagger courage as the white-shrouded bear or shark.[1]

---

[1] With reference to the Polar bear, it is not the whiteness, separately regarded, which heightens the intolerable hideousness of that brute; for, analysed, that heightened hideousness only arises from the circumstance, that the irresponsible ferociousness of the creature stands invested in the fleece of celestial innocence and love; and hence, by bringing together two such opposite emotions in our minds, the Polar bear frightens us with so unnatural a contrast. But even assuming all this to be true; yet, were it not for the whiteness, you would not have that intensified terror.

"Moby-Dick," 1984. *"Thus, the muffled rollings of a milky sea; the bleak rustlings of the festooned frosts of mountains; the desolate shiftings of the windrowed snows of prairies; all these, to Ishmael, are as the shaking of the buffalo robe to the frightened colt! Though neither knows where lie the nameless things of which the mystic sign gives forth such hints; yet with me, as with the colt, somewhere those things must exist. Though in many of its aspects this visible world seems formed in love, the invisible spheres were formed in fright." — Chapter 42. Mural painting by Richard Ellis at New Bedford Whaling Museum.*

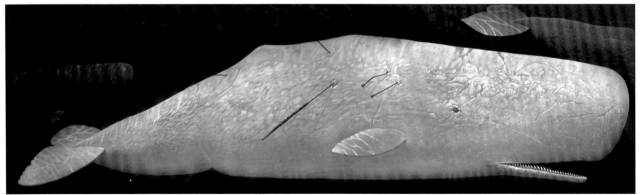

New Bedford Whaling Museum

What is it that in the Albino man so peculiarly repels and often shocks the eye, as that sometimes he is loathed by his own kith and kin! It is that whiteness which invests him, a thing expressed by the name he bears. The Albino is as well made as other men—has no substantive deformity—and yet this mere aspect of all-pervading whiteness makes him more strangely hideous than the ugliest abortion. Why should this be so?

Nor, in some things, does the common, hereditary experience of all mankind fail to bear witness to the supernaturalism of this hue. It cannot well be doubted, that the one visible quality in the aspect of the dead which most appals the gazer, is the marble pallor lingering there; as if indeed that pallor were as much like the badge of consternation in the other world, as of mortal trepidation here. And from that pallor of the dead, we borrow the expressive hue of the shroud in which we wrap them. Nor even in our superstitions do we fail to throw the same snowy mantle round our phantoms; all ghosts rising in a milk-white fog.

Therefore, in his other moods, symbolize whatever grand or gracious thing he will by whiteness, no man can deny that in its profoundest idealized significance it calls up a peculiar apparition to the soul.

But though without dissent this point be fixed, how is mortal man to account for it? To analyse it, would seem impossible. Can we thus hope to light upon some chance clue to conduct us to the hidden cause we seek?

I know that, to the common apprehension, this phenomenon of whiteness is not confessed to be the prime agent in exaggerating the terror of objects otherwise terrible.

The mariner, when drawing nigh the coasts of foreign lands, if by night he hear the roar of breakers, starts to vigilance, and feels just enough of trepidation to sharpen all his faculties; but under precisely similar circumstances, let him be called from his hammock to view his ship sailing through a midnight sea of milky whiteness—as if from encircling headlands shoals of combed white bears were swimming round him, then he feels a silent, superstitious dread; the shrouded phantom of the whitened waters is horrible to him as a real ghost; in vain the lead assures him he is still off soundings; heart and helm they both go down; he never rests till blue water is under him again. Yet where is the mariner who will tell thee, "Sir, it was not so much the fear of striking hidden rocks, as the fear of that hideous whiteness that so stirred me?"

To the native Indian of Peru, the continual sight of the snow-howdahed Andes conveys naught of dread, except, perhaps, in the mere fancying of the eternal frosted desolateness reigning at such vast altitudes, and the natural conceit of what a fearfulness it would be to lose oneself in such inhuman solitudes. Much the same is it with the backwoodsman of the West, who with comparative indifference views an unbounded prairie sheeted with driven snow, no shadow of tree or twig to break the fixed trance of whiteness. Not so the sailor, beholding the scenery of the Antarctic seas; where at times, by some infernal trick of legerdemain in the powers of frost and air, he, shivering and half shipwrecked, instead of rainbows speaking hope and solace to his misery, views what seems a boundless church-yard grinning upon him with its lean ice monuments and splintered crosses.

But thou sayest, methinks this white-lead chapter about whiteness is but a white flag hung out from a craven soul; thou surrenderest to a hypo, Ishmael.

But not yet have we solved the incantation of this whiteness, and learned why it appeals with such power to the soul; and more strange and far more portentous—why, as we have seen, it is at once the most meaning symbol of spiritual things, nay, the very veil of the Christian's Deity; and yet should be as it is, the intensifying agent in things the most appalling to mankind.

Is it that by its indefiniteness it shadows forth the heartless voids and immensities of the universe, and thus stabs us from behind with the thought of annihilation, when beholding the white depths of the milky way? Or is it, that as in essence whiteness is not so much a color as the visible absence of color, and at the same time the concrete of all colors; is it for these reasons that there is such a dumb blankness, full of meaning, in a wide landscape of snows—a colorless, all- color of atheism from which we shrink? And when we consider that other theory of the natural philosophers, that all other earthly hues—every stately or lovely emblazoning—the sweet tinges of sunset skies and woods; yea, and the gilded velvets of butterflies, and the butterfly cheeks of young girls; all these are but subtile deceits, not actually inherent in substances, but only laid on from without; and consider that the mystical cosmetic which produces every one of her hues, the great principle of light, for ever remains white or colorless in itself, and if operating without medium upon matter, would touch all objects, even tulips and roses, with its own blank tinge—pondering all this, the palsied universe lies before us a leper; and like wilful travellers in Lapland, who refuse to wear colored and coloring glasses upon their eyes, so the wretched infidel gazes himself blind at the monumental white shroud that wraps all the prospect around him. And of all these things the Albino Whale was the symbol. Wonder ye then at the fiery hunt?

Chapter XLIII, XLIV

Hark! · The Chart

Had you followed Captain Ahab down into his cabin after the squall that took place on the night succeeding that wild ratification of his purpose with his crew, you would have seen him go to a locker in the transom, and bringing out a large wrinkled roll of yellowish sea charts, spread them before him on his screwed-down table. Then seating himself before it, you would have seen him intently study the various lines and shadings which there met his eye; and with slow but steady pencil trace additional courses over spaces that before were blank. At intervals, he would refer to piles of old log-books beside him, wherein were set down the seasons and places in which, on various former voyages of various ships, sperm whales had been captured or seen.

Maury Whale Chart, 1851. *This invaluable 19th century resource was the whalers' bible and the last word on where to find the best fishing grounds for sperm and right whales. Using small icons of whales, the chart provided data on the species' populations and locations, the seasons and the frequency that whales visited areas, and how they traveled (in pairs, schools or straggling). It was updated and revised regularly.* — By Lt. Matthew F. Maury, United States Navy, Washington National Observatory.

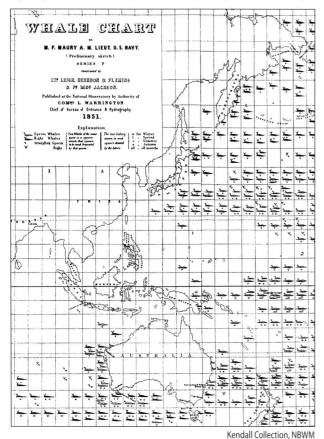

Kendall Collection, NBWM

"The Navigator." —Oil painting by Clifford W. Ashley. New Bedford Whaling Museum.

Now, to any one not fully acquainted with the ways of the leviathans, it might seem an absurdly hopeless task thus to seek out one solitary creature in the unhooped oceans of this planet. But not so did it seem to Ahab, who knew the sets of all tides and currents; and thereby calculating the driftings of the sperm whale's food; and, also, calling to mind the regular, ascertained seasons for hunting him in particular latitudes; could arrive at reasonable surmises, almost approaching to certainties, concerning the timeliest day to be upon this or that ground in search of his prey.

So assured, indeed, is the fact concerning the periodicalness of the sperm whale's resorting to given waters, that many hunters believe that, could he be closely observed and studied throughout the world; were the logs for one voyage of the entire whale fleet carefully collated, then the migrations of the sperm whale would be found to correspond in invariability to those of the herring-shoals or the flights of swallows. On this hint, attempts have been made to construct elaborate migratory charts of the sperm whale.

Note: Since the above was written, the statement is happily borne out by an official circular, issued by Lieutenant Maury, of the National Observatory, Washington, April 16th, 1851. By that circular, it appears that precisely such a chart is in course of completion; and portions of it are presented in the circular. "This chart divides the ocean into districts of five degrees of latitude by five degrees of longitude; perpendicularly through each of which districts are twelve columns for the twelve months; and horizontally through each of which districts are three lines; one to show the number of days that have been spent in each month in every district, and the two others to show the number of days in which whales, sperm or right, have been seen."

Chapter XLV

The Affadavit

I do not know where I can find a better place than just here, to make mention of one or two other things, which to me seem important. So ignorant are most landsmen of some of the plainest and most palpable wonders of the world, that without some hints touching the plain facts, historical and otherwise, of the fishery, they might scout at Moby Dick as a monstrous fable, or still worse and more detestable, a hideous and intolerable allegory.

First: Though most men have some vague flitting ideas of the general perils of the grand fishery, yet they have nothing like a fixed, vivid conception of those perils, and the frequency with which they recur. One reason perhaps is, that not one in fifty of the actual disasters and deaths by casualties in the fishery, ever finds a public record at home, however transient and immediately forgotten that record. Yet I tell you that upon one particular voyage which I made to the Pacific, among many others we spoke thirty different ships, every one of which had had a death by a whale, some of them more than one, and three that had each lost a boat's crew. For God's sake, be economical with your lamps and candles! not a gallon you burn, but at least one drop of man's blood was spilled for it.

Secondly: People ashore have indeed some indefinite idea that a whale is an enormous creature of enormous power.

The sperm whale is in some cases sufficiently powerful, knowing, and judiciously malicious, as with direct aforethought to stave in, utterly destroy, and sink a large ship; and what is more, the sperm whale has done it.

In the year 1820 the ship Essex, Captain Pollard, of Nantucket, was cruising in the Pacific Ocean. One day she saw spouts, lowered her boats, and gave chase to a shoal of sperm whales. Ere long, several of the whales were wounded; when, suddenly, a very large whale escaping from the boats, issued from the shoal, and bore directly down upon the ship. Dashing his forehead against her hull, he so stove her in, that in less than "ten minutes" she settled down and fell over. Not a surviving plank of her has been seen since. After the severest exposure, part of the crew reached the land in their boats. Being returned home at last, Captain Pollard once more sailed for the Pacific in command of another ship, but the gods shipwrecked him again upon unknown rocks and breakers; for the second time his ship was utterly lost, and forthwith forswearing the sea, he has never tempted it since.[1]

Survivors of the Essex rescued after 90 days at sea. *"When I was on board the* Acushnet *of Fairhaven, on the passage to the Pacific cruising grounds, among other matters of forecastle conversation at times was the story of the* Essex...*and her truly astounding fate.... Reading of this wondrous story upon the landless sea, and close to the very latitude of the shipwreck had a surprising effect upon me."* — Melville's memoir of Owen Chase. From Leyda, The Melville Log. Illustration from Murphy's Journal. Kendall Collection, NBWM.

The ship Union, also of Nantucket, was in the year 1807 totally lost off the Azores by a similar onset.

Some eighteen or twenty years ago Commodore J— then commanding an American sloop-of-war of the first class, happened to be dining with a party of whaling captains, on board a Nantucket ship in the harbor of Oahu, Sandwich Islands. Conversation turning upon whales, the Commodore was pleased to be sceptical touching the amazing strength ascribed to them by the professional gentlemen present. He peremptorily denied for example, that any whale could so smite his stout sloop-of-war as to cause her to leak so much as a thimbleful. Very good; but there is more coming. Some weeks after, the commodore set sail in this impregnable craft for Valparaiso. But he was stopped on the way by a portly sperm whale, that begged a few moments' confidential business with him. That business consisted in fetching the Commodore's craft such a thwack, that with all his pumps going he made straight for the nearest port to heave down and repair. I am not superstitious, but I consider the Commodore's interview with that whale as providential. Was not Saul of Tarsus converted from unbelief by a similar fright? I tell you, the sperm whale will stand no nonsense.

---

[1] The following [is] from Chase's narrative: "He made two several attacks upon the ship, both of which, according to their direction, were calculated to do us the most injury....His aspect was most horrible, and such as indicated resentment and fury. He came directly from the shoal which we had just before entered, and in which we had struck three of his companions, as if fired with revenge for their sufferings....The dark ocean and swelling waters were nothing; the fears of being swallowed up by some dreadful tempest, or dashed upon hidden rocks... seemed scarcely entitled to a moment's thought; the dismal looking wreck, and the horrid aspect and revenge of the whale, wholly engrossed my reflections, until day again made its appearance."

Chapter XLVII

The Mat Maker

Whaleboat on the *John Manta*, 1925. — *William R. Hegarty Collection*

It was a cloudy, sultry afternoon; the seamen were lazily lounging about the decks, or vacantly gazing over into the lead-colored waters. Queequeg and I were mildly employed weaving what is called a sword-mat, for an additional lashing to our boat.

I was the attendant or page of Queequeg, while busy at the mat. As I kept passing and repassing the filling or woof of marline between the long yarns of the warp, using my own hand for the shuttle, and as Queequeg, standing sideways, ever and anon slid his heavy oaken sword between the threads, and idly looking off upon the water, carelessly and unthinkingly drove home every yarn: so strange a dreaminess did there then reign all over the ship and all over the sea, that it seemed as if this were the Loom of Time, and I myself were a shuttle mechanically weaving and weaving away at the Fates. Here, thought I, with my own hand I ply my own shuttle and weave my own destiny into these unalterable threads.

Thus we were weaving and weaving away when I started at a sound so strange, long drawn, and musically wild and unearthly, that the ball dropped from my hand, and I stood gazing up at the clouds whence that voice dropped like a wing. High aloft in the cross-trees was that mad Gay-Header, Tashtego. His body was reaching eagerly forward, his hand stretched out like a wand, and at brief sudden intervals he continued his cries.

"There she blows! there! there! there! she blows! she blows!"

"Where-away?"

"On the lee-beam, about two miles off! a school of them!"

Instantly all was commotion.

The Sperm Whale blows as a clock ticks, with the same undeviating and reliable uniformity. And thereby whalemen distinguish this fish from other tribes of his genus.

"There go flukes!" was now the cry from Tashtego; and the whales disappeared.

"Quick, steward!" cried Ahab. "Time! time!"

Dough-Boy hurried below, glanced at the watch, and reported the exact minute to Ahab.

One of the men selected for shipkeepers—that is, those not appointed to the boats, relieved the Indian at the main-mast head. The sailors at the fore and mizzen had come down; the line tubs were fixed in their places; the cranes were thrust out; the mainyard was backed, and the three boats swung over the sea like three samphire baskets over high cliffs. Outside of the bulwarks their eager crews with one hand clung to the rail, while one foot was expectantly poised on the gunwale. So look the long line of man-of-war's men about to throw themselves on board an enemy's ship.

But at this critical instant a sudden exclamation was heard that took every eye from the whale. With a start all glared at dark Ahab, who was surrounded by five dusky phantoms that seemed fresh formed out of air.

Lower Away! *Whaleboat ready to go on the* John R. Manta *awaits the chase, 1925.*

The captain's boat is the first to be lowered on the brig Daisy, *1912.*

The chase begins as whaleboat and crew from the John R. Manta *hit the water, 1925.*

William R. Hegarty Collection

New Bedford Whaling Museum

William R. Hegarty Collection

CHAPTER XLVIII

The First Lowering

The phantoms, for so they then seemed, were flitting on the other side of the deck, and, with a noiseless celerity, were casting loose the tackles and bands of the boat which swung there. This boat had always been deemed one of the spare boats, though technically called the captain's, on account of its hanging from the starboard quarter. The figure that now stood by its bows was tall and swart, with one white tooth evilly protruding from its steel-like lips. A rumpled Chinese jacket of black cotton funereally invested him, with wide black trowsers of the same dark stuff. But strangely crowning his ebonness was a glistening white plaited turban, the living hair braided and coiled round and round upon his head. Less swart in aspect, the companions of this figure were of that vivid, tiger-yellow complexion peculiar to some of the aboriginal natives of the Manillas;—a race notorious for a certain diabolism of subtilty, and by some honest white mariners supposed to be the paid spies and secret confidential agents on the water of the devil, their lord, whose counting-room they suppose to be elsewhere.

While yet the wondering ship's company were gazing upon these strangers, Ahab cried out to the white-turbaned old man at their head, "All ready there, Fedallah?"

"Ready," was the half-hissed reply.

"Lower away then; d'ye hear?" shouting across the deck. "Lower away there, I say."

Such was the thunder of his voice, that spite of their amazement the men sprang over the rail; the sheaves whirled round in the blocks; with a wallow, the three boats dropped into the sea; while, with a dexterous, off-handed daring, unknown in any other vocation, the sailors, goat-like, leaped down the rolling ship's side into the tossed boats below.

Hardly had they pulled out from under the ship's lee, when a fourth keel, coming from the windward

Oarsmen "pulling to" on the Daisy, *1912. The iron is in position for easy access to the boasteerer who will soon become harpooner.* "With the still rising wind, we rushed along; the boat going with such madness through the water, that the lee oars could scarcely be worked rapidly enough to escape being torn from the row-locks." *— Chapter 48. Robert Cushman Murphy photograph. New Bedford Whaling Museum.*

side, pulled round under the stern, and showed the five strangers rowing Ahab, who, standing erect in the stern, loudly hailed Starbuck, Stubb, and Flask, to spread themselves widely, so as to cover a large expanse of water. But with all their eyes again riveted upon the swart Fedallah and his crew, the inmates of the other boats obeyed not the command.

"Spread yourselves," cried Ahab; "give way, all four boats. Thou, Flask, pull out more to leeward!"

"Aye, aye, sir," cheerily cried little King-Post, sweeping round his great steering oar.

"Pull, pull, my fine hearts-alive; pull, my children; pull, my little ones," drawingly and soothingly sighed Stubb to his crew, some of whom still showed signs of uneasiness. "Why don't you break your backbones, my boys? What is it you stare at? Those chaps in yonder boat? Tut! They are only five more hands come to help us—never mind from where—the more the merrier. Pull, then, do pull; never mind the brimstone—devils

Crew of the Charles W. Morgan, *1912. After pulling alongside the whale, the harpooner gets ready to strike.* "Stubbs would say the most terrific things to his crew, in a tone so strangely compounded of fun and fury, and the fury seemed so calculated merely as a spice to the fun, that no oarsman could hear such queer invocations without pulling for dear life, and yet pulling for the mere joke of the thing." *— Chapter 48.*

Edwards Family Collection

are good fellows enough. So, so; there you are now; that's the stroke for a thousand pounds; that's the stroke to sweep the stakes! Hurrah for the gold cup of sperm oil, my heroes! Three cheers, men—all hearts alive! Easy, easy; don't be in a hurry—don't be in a hurry. Why don't you snap your oars, you rascals? Bite something, you dogs! So, so, so, then;—softly, softly! That's it—that's it! long and strong. Give way there, give way! The devil fetch ye, ye ragamuffin rapscallions; ye are all asleep. Stop snoring, ye sleepers, and pull. Pull, will ye? pull, can't ye? pull, won't ye? Why in the name of gudgeons and ginger-cakes don't ye pull?—pull and break something! pull, and start your eyes out! Here!" whipping out the sharp knife from his girdle; "every mother's son of ye draw his knife, and pull with the blade between his teeth. That's it—that's it. Now ye do something; that looks like it, my steel-bits. Start her—start her, my silver-spoons! Start her, marling-spikes!"

In obedience to a sign from Ahab, Starbuck was now pulling obliquely across Stubb's bow; and when for a minute or so the two boats were pretty near to each other, Stubb hailed the mate.

"Mr. Starbuck! Larboard boat there, ahoy! A word with ye, sir, if ye please!"

"Halloa!" returned Starbuck, turning round not a single inch as he spoke; still earnestly but whisperingly urging his crew; his face set like a flint from Stubb's.

"What think ye of those yellow boys, sir!"

"Smuggled on board, somehow, before the ship sailed. A sad business, Mr. Stubb! But never mind, Mr. Stubb, all for the best. There's hogsheads of sperm ahead, Mr. Stubb, and that's what ye came for. Sperm, sperm's the play! This at least is duty; duty and profit hand in hand!"

"Aye, aye, I thought as much," soliloquized Stubb, when the boats diverged, "as soon as I clapt eye on 'em, I thought so. Aye, and that's what he went into the after hold so often for. They were hidden down there. The white whale's at the bottom of it. Well, well, so be it! Can't be helped!"

Meantime, Ahab, having sided the furthest to windward, was still ranging ahead of the other boats; a circumstance bespeaking how potent a crew was pulling him. Those tiger yellow creatures of his seemed all steel and whale-bone; like five trip-hammers they rose and fell with regular strokes of strength, which periodically started the boat along the water like a horizontal burst boiler out of a Mississippi steamer.

But what it was that inscrutable Ahab said to that tiger-yellow crew of his—these were words best omitted here; for you live under the blessed light of the evangelical land. Only the infidel sharks in the audacious seas may give ear to such words, when, with tornado brow, and eyes of red murder, and foam-glued lips, Ahab leaped after his prey.

It was a sight full of quick wonder and awe! The vast swells of the omnipotent sea; the surging, hollow roar they made, as they rolled along the eight gunwales, like gigantic bowls in a boundless bowling-green; the brief suspended agony of the boat, as it would tip for

Ready to strike, 1922. In a scene from the motion picture, "Down to the Sea in Ships," our hero is set to strike a sperm whale. "Not the raw recruit, marching from the bosom of his wife into the fever heat of his first battle; not the dead man's ghost encountering the first unknown phantom in the other world;—neither of these can feel stranger and stronger emotions than that man does, who for the first time finds himself pulling into the charmed, churned circle of the hunted Sperm Whale." — *Chapter 48.*

New Bedford Whaling Museum

"Struck on a breach," 1876. *"Though not one of the oarsmen was then facing the life and death peril so close to them ahead, yet with their eyes on the intense countenance of the mate in the stern of the boat, they knew that the imminent instant had come; they heard, too, an enormous wallowing sound as of fifty elephants stirring in their litter. Meanwhile the boat was still booming through the mist, the waves curling and hissing around us like the erected crests of enraged serpents."* — Chapter 48. Engraving from Davis: Nimrod of the Sea.

Stove boat, circa 1872. — Ink and pencil drawing by E. C. Snow, aboard Abraham Baker. — Kendall Collection, NBWM.

an instant on the knife-like edge of the sharper waves, that almost seemed threatening to cut it in two; the sudden profound dip into the watery glens and hollows; the keen spurrings and goadings to gain the top of the opposite hill; the headlong, sled-like slide down its other side;—all these, with the cries of the headsmen and harpooneers, and the shuddering gasps of the oarsmen, with the wondrous sight of the ivory Pequod bearing down upon her boats with outstretched sails, like a wild hen after her screaming brood;—all this was thrilling.

Soon after, two cries in quick succession on each side of us denoted that the other boats had got fast; but hardly were they overheard, when with a lightning-like hurtling whisper Starbuck said: "Stand up!" and Queequeg, harpoon in hand, sprang to his feet.

"That's his hump. *There, there*, give it to him!" whispered Starbuck.

A short rushing sound leaped out of the boat; it was the darted iron of Queequeg. Then all in one welded commotion came an invisible push from astern, while forward the boat seemed striking on a ledge; the sail collapsed and exploded; a gush of scalding vapor shot up near by; something rolled and tumbled like an earthquake beneath us. The whole crew were half suffocated as they were tossed helter-skelter into the white curdling cream of the squall. Squall, whale, and harpoon had all blended together; and the whale, merely grazed by the iron, escaped.

Though completely swamped, the boat was nearly unharmed. Swimming round it we picked up the floating oars, and lashing them across the gunwale, tumbled back to our places. There we sat up to our knees in the sea, the water covering every rib and plank, so that to our downward gazing eyes the suspended craft seemed a coral boat grown up to us from the bottom of the ocean.

The wind increased to a howl; the waves dashed their bucklers together; the whole squall roared, forked, and crackled around us like a white fire upon the prairie, in which, unconsumed, we were burning; immortal in these jaws of death! The rising sea forbade all attempts to bale out the boat. The oars were useless as propellers, performing now the office of life-preservers. So, cutting the lashing of the water-proof match keg, after many failures Starbuck contrived to ignite the lamp in the lantern; then stretching it on a waif pole, handed it to Queequeg as the standard-bearer of this forlorn hope. There, then, he sat, holding up that imbecile candle in the heart of that almighty forlornness. There, then, he sat, the sign and symbol of a man without faith, hopelessly holding up hope in the midst of despair.

Wet, drenched through, and shivering cold, despairing of ship or boat, we lifted up our eyes as the dawn came on. Suddenly Queequeg started to his feet, hollowing his hand to his ear. We all heard a faint creaking, as of ropes and yards hitherto muffled by the storm. Affrighted, we all sprang into the sea as the ship at last loomed into view, bearing right down upon us within a distance of not much more than its length.

Floating on the waves we saw the abandoned boat, as for one instant it tossed and gaped beneath the ship's bows like a chip at the base of a cataract; and then the vast hull rolled over it, and it was seen no more till it came up weltering astern. Again we swam for it, were dashed against it by the seas, and were at last taken up and safely landed on board. Ere the squall came close to, the other boats had cut loose from their fish and returned to the ship in good time. The ship had given us up, but was still cruising, if haply it might light upon some token of our perishing,—an oar or a lance pole.

Chapter XLIX

The Hyena

"He came up alongside of the Boat, and turned it over with his Nose as a Hog would his Eating-trough." — *Bullen*, Cruise of the Cachalot.

There are certain queer times and occasions in this strange mixed affair we call life when a man takes this whole universe for a vast practical joke, though the wit thereof he but dimly discerns, and more than suspects that the joke is at nobody's expense but his own. There is nothing like the perils of whaling to breed this free and easy sort of genial, desperado philosophy; and with it I now regarded this whole voyage of the Pequod, and the great white whale its object.

Considering, therefore, that squalls and capsizings in the water and consequent bivouacks on the deep, were matters of common occurrence in this kind of life; considering that at the superlatively critical instant of going on to the whale I must resign my life into the hands of him who steered the boat—oftentimes a fellow who at that very moment is in his impetuousness upon the point of scuttling the craft with his own frantic stampings; considering that the particular disaster to our own particular boat was chiefly to be imputed to Starbuck's driving on to his whale almost in the teeth of a squall, and considering that Starbuck, notwithstanding, was famous for his great heedfulness in the fishery; considering that I belonged to this uncommonly prudent Starbuck's boat; and finally considering in what a devil's chase I was implicated, touching the White Whale: taking all things together, I say, I thought I might as well go below and make a rough draft of my will. "Queequeg," said I, "come along, you shall be my lawyer, executor, and legatee."

It may seem strange that of all men sailors should be tinkering at their last wills and testaments, but there are no people in the world more fond of that diversion. This was the fourth time in my nautical life that I had done the same thing. After the ceremony was concluded upon the present occasion, I felt all the easier; a stone was rolled away from my heart.

Now then, thought I, unconsciously rolling up the sleeves of my frock, here goes a cool, collected dive at death and destruction, and the devil fetch the hindmost.

"Taking a Whale," circa 1865. — *Drawing by Robert W. Weir, Jr.*

Kendall Collection, NBWM

"Greenland Whale," 1837. Mocking the whaleman's sublime delight in confronting terror, Ishmael remarks: "I suppose then, that going plump on a flying whale with your sail set in a foggy squall is the height of a whaleman's discretion?"

But such mockery, and humor, enables these whalemen to fly fearless into this world of terror. "And as for small difficulties and worryings, prospects of sudden disaster, peril of life and limb; all these, and death itself, seem only sly, good-natured hits… That odd sort of wayward mood I am speaking of, comes over a man only in some time of extreme tribulation; it comes in the very midst of his earnestness, so that what just before might have seemed to him a thing most momentous, now seems but a part of the general joke…" — Chapter 49. *Colored lithograph from Hamilton:* The Natural History of…Whales. *Drawing by J. Stewart, engraved by W. Lizars.*

New Bedford Whaling Museum

"Shooting a whale with a shoulder gun." This painting by Robert Wallace Weir, Jr. (1836-1905), as well as the woodcut illustration on opposite page, is based on a drawing from his journal aboard the bark *Clara Bell* (1855-58). Weir was the black sheep of an emerging dynasty of American artists. He ran away to sea at age 15 and used "Robert Wallace" as a pseudonym on three of his four whaling voyages (1851-62).

Kendall Collection, NBWM

Chapter L

Ahab's boat and crew. Fedallah

Fayal, Azores, 1841. — *From journal of ship* Lucy Ann. *Kendall Collection, NBWM.*

Ahab well knew that although his friends at home would think little of his entering a boat in certain comparatively harmless vicissitudes of the chase, for the sake of being near the scene of the action and giving his orders in person, yet for Captain Ahab to have a boat actually apportioned to him as a regular headsman in the hunt—above all for Captain Ahab to be supplied with five extra men, as that same boat's crew, he well knew that such generous conceits never entered the heads of the owners of the Pequod. Therefore he had not solicited a boat's crew from them, nor had he in any way hinted his desires on that head. Nevertheless he had taken private measures of his own touching all that matter.

Certain it is that while the subordinate phantoms soon found their place among the crew, though still as it were somehow distinct from them, yet that Fedallah remained a muffled mystery to the last.

Whence he came in a mannerly world like this, by what sort of accountable tie he soon evinced himself to be linked with Ahab's peculiar fortunes; nay, so far as to have some sort of a half-hinted influence; Heaven knows. But one cannot sustain an indifferent air concerning Fedallah. He was such a creature as civilized, domestic people in the temperate zone only see in their dreams, and that but dimly; but the like of whom now and then glide among the unchanging Asiatic communities, especially the Oriental isles to the east of the continent.

"Port of Horta, Fayal and Pico of the The Azores Islands," 1945. *Among the whaleships in port at Horta is the Charles W. Morgan (just right of center). The Azores and the Cape Verde Islands were the first two ports of call for most New England whaleships stopping for provisions and crew en route to the sperm whaling area known as the Western Grounds.*

Chapter LI
The Spirit Spout

São Jaoa, Cape Verde Islands, 1867.
— By R.G.N. Swift, aboard ship Contest. Kendall Collection, NBWM.

Days, weeks passed, and under easy sail, the ivory Pequod had slowly swept across four several cruising-grounds; that off the Azores; off the Cape de Verdes; on the Plate (so called), being off the mouth of the Rio de la Plata; and the Carrol Ground, an unstaked, watery locality, southerly from St. Helena.

But, at last, when turning to the eastward, the Cape winds began howling around us, and we rose and fell upon the long, troubled seas that are there; when the ivory-tusked Pequod sharply bowed to the blast, and gored the dark waves in her madness, till, like showers of silver chips, the foam-flakes flew over her bulwarks; then all this desolate vacuity of life went away, but gave place to sights more dismal than before.

Close to our bows, strange forms in the water darted hither and thither before us; while thick in our rear flew the inscrutable sea-ravens. And every morning, perched on our stays, rows of these birds were seen; and spite of our hootings, for a long time obstinately clung to the hemp, as though they deemed our ship some drifting, uninhabited craft; a thing appointed to desolation, and therefore fit roosting-place for their homeless selves. And heaved and heaved, still unrestingly heaved the black sea, as if its vast tides were a conscience; and the great mundane soul were in anguish and remorse for the long sin and suffering it had bred.

The Cape Verde Islands (bottom), Brava and Fago (with volcano), 1845. *Long after the whaling industry died, Portuguese and Cape Verdean families continued to migrate to southeastern New England. Today the region is home to the largest and oldest Portuguese and Cape Verdean communities in America.* — Russell and Purrington's Panorama

New Bedford Whaling Museum

The Cape of Good Hope, 1794. *French and Spanish vessels plying Cape waters.* —By J. Van Ryne, pub. by Laurie & Whittle, London.

Cape of Good Hope, do they call ye? Rather Cape Tormentoto, as called of yore; for long allured by the perfidious silences that before had attended us, we found ourselves launched into this tormented sea, where guilty beings transformed into those fowls and these fish, seemed condemned to swim on everlastingly without any haven in store, or beat that black air without any horizon. But calm, snow-white, and unvarying; still directing its fountain of feathers to the sky; still beckoning us on from before, the solitary jet would at times be descried.

During all this blackness of the elements, Ahab, though assuming for the time the almost continual command of the drenched and dangerous deck, manifested the gloomiest reserve; and more seldom than ever addressed his mates. In tempestuous times like these, after everything above and aloft has been secured, nothing more can be done but passively to await the issue of the gale. Then Captain and crew become practical fatalists. So, with his ivory leg inserted into its accustomed hole, and with one hand firmly grasping a shroud, Ahab for hours and hours would stand gazing dead to windward, while an occasional squall of sleet or snow would all but congeal his very eyelashes together. Meantime, the crew driven from the forward part of the ship by the perilous seas that burstingly broke over its bows, stood in a line along the bulwarks in the waist; and the better to guard against the leaping waves, each man had slipped himself into a sort of bowline secured to the rail, in which he swung as in a loosened belt. Few or no words were spoken; and the silent ship, as if manned by painted sailors in wax, day after day tore on through all the swift madness and gladness of the demoniac waves. By night the same muteness of humanity before the shrieks of the ocean prevailed; still in silence the men swung in the bowlines; still wordless Ahab stood up to the blast. Even when wearied nature seemed demanding repose he would not seek that repose in his hammock. Never could Starbuck forget the old man's aspect, when one night going down into the cabin to mark how the barometer stood, he saw him with closed eyes sitting straight in his floor-screwed chair; the rain and half-melted sleet of the storm from which he had some time before emerged, still slowly dripping from the unremoved hat and coat. On the table beside him lay unrolled one of those charts of tides and currents which have previously been spoken of. His lantern swung from his tightly clenched hand. Though the body was erect, the head was thrown back so that the closed eyes were pointed towards the needle of the tell-tale that swung from a beam in the ceiling.

Terrible old man! thought Starbuck with a shudder, sleeping in this gale, still thou steadfastly eyest thy purpose.

Ships in moonlight, Atlantic Ocean, 1845. —Russell and Purrington's Panorama

Kendall Collection, NBWM

100

Chapter LII

The Albatross

Logbook art by John C. Scales aboard bark *Pearl* of New London, 1852-54. — *Kendall Collection, NBWM*

South-eastward from the Cape, off the distant Crozetts, a good cruising ground for right whalemen, a sail loomed ahead, the Goney (Albatross) by name. As she slowly drew nigh, from my lofty perch at the foremast-head, I had a good view of that sight so remarkable to a tyro in the far ocean fisheries—a whaler at sea, and long absent from home.

A wild sight it was to see her long-bearded lookouts at those three mast-heads. Standing in iron hoops nailed to the mast, they swayed and swung over a fathomless sea; and though, when the ship slowly glided close under our stern, we six men in the air came so nigh to each other that we might almost have leaped from the mast-heads of one ship to those of the other; yet, those forlorn-looking fishermen, mildly eyeing us as they passed, said not one word to our own look-outs, while the quarter-deck hail was being heard from below.

"Ship ahoy! Have ye seen the white whale?"

But as the strange captain, leaning over the pallid bulwarks, was in the act of putting his trumpet to his mouth, it somehow fell from his hand into the sea; and the wind now rising amain, he in vain strove to make himself heard without it. Meantime, his ship was still increasing the distance between.

Ahab for a moment paused; it almost seemed as though he would have lowered a boat to board the stranger, had not the threatening wind forbade. But taking advantage of his windward position, he again seized his trumpet, and knowing by her aspect that the stranger vessel was a Nantucketer and shortly bound home, he loudly hailed—"Ahoy there! This is the Pequod, bound round the world! Tell them to address all future letters to the Pacific Ocean! And this time three years, if I am not home, tell them to address them to—"

At that moment the two wakes were fairly crossed, and instantly, then, shoals of small harmless fish, that for some days before had been placidly swimming by our side, darted away with what seemed shuddering fins, and ranged themselves fore and aft with the stranger's flanks.

"Swim away from me, do ye?" murmured Ahab, gazing over into the water. There seemed little in the words, but the tone conveyed more of deep helpless sadness than the insane old man had ever before evinced. But turning to the steersman, who thus far had been holding the ship in the wind to diminish her headway, he cried out in his old lion voice,—"Up helm! Keep her off round the world!"

Bark Arab of Fairhaven, 1837-38. *The* Arab *passes friendly territory, heading to the South Atlantic and Indian Ocean grounds.* — *Anonymous artist.*

New Bedford Whaling Museum

Chapter LIII

The Gam

"The Christmas Mail." — by Charles S. Raleigh. New Bedford Whaling Museum.

The ostensible reason why Ahab did not go on board of the whaler we had spoken was this: the wind and sea betokened storms. But even had this not been the case, he would not after all, perhaps, have boarded her—judging by his subsequent conduct on similar occasions—if so it had been that, by the process of hailing, he had obtained a negative answer to the question he put. For, as it eventually turned out, he cared not to consort, even for five minutes, with any stranger captain, except he could contribute some of that information he so absorbingly sought. But all this might remain inadequately estimated, were not something said here of the peculiar usages of whaling-vessels when meeting each other in foreign seas, and especially on a common cruising-ground.

If two strangers crossing the Pine Barrens in New York State, or the equally desolate Salisbury Plain in England; if casually encountering each other in such inhospitable wilds, these twain, for the life of them, cannot well avoid a mutual salutation; and stopping for a moment to interchange the news; and, perhaps, sitting down for a while and resting in concert: then, how much more natural that upon the illimitable Pine Barrens and Salisbury Plains of the sea, two whaling vessels descrying each other at the ends of the earth—off lone Fanning's Island, or the far away King's Mills; how much more natural, I say, that under such circumstances these ships should not only interchange hails, but come into still closer, more friendly and sociable contact. And especially would this seem to be a matter of course, in the case of vessels owned in one seaport, and whose captains, officers, and not a few of the men are personally known

"Having a 'Gam'—No Whales in Sight," circa 1865. *"The whale-boat has no seat astern and no tiller at all; and therefore a complete boat's crew must leave the ship, and the captain, having no place to sit in, is pulled off to his visit all standing like a pine tree. And this standing captain is all alive to the importance of sustaining his dignity by maintaining his legs. It would never do in plain sight of the world's riveted eyes, for this straddling captain to be seen steadying himself the slightest particle by catching hold of anything with his hands. Nevertheless there have occurred instances, where the captain has been known for an uncommonly critical moment or two, in a sudden squall say—to seize hold of the nearest oarsman's hair, and hold on there like grim death."* —Chapter 53. Watercolor by Benjamin Russell.

New Bedford Whaling Museum

to each other; and consequently, have all sorts of dear domestic things to talk about.

For the long absent ship, the outward-bounder, perhaps, has letters on board; at any rate, she will be sure to let her have some papers of a date a year or two later than the last one on her blurred and thumb-worn files. And in return for that courtesy, the outward-bound ship would receive the latest whaling intelligence from the cruising-ground to which she may be destined, a thing of the utmost importance to her. And in degree, all this will hold true concerning whaling vessels crossing each other's track on the cruising-ground itself, even though they are equally long absent from home. For one of them may have received a transfer of letters from some third, and now far remote vessel; and some of those letters may be for the people of the ship she now meets. Besides, they would exchange the whaling news, and have an agreeable chat. For not only would they meet with all the sympathies of sailors, but likewise with all the peculiar congenialities arising from a common pursuit and mutually shared privations and perils.

Nor would difference of country make any very essential difference; that is, so long as both parties speak one language, as is the case with Americans and English. Though, to be sure, from the small number of English whalers, such meetings do not very often occur, and when they do occur there is too apt to be a sort of shyness between them; for your Englishman is rather reserved, and your Yankee, he does not fancy that sort of thing in anybody but himself.

Of all ships separately sailing the sea, the whalers have most reason to be sociable—and they are so. What does the whaler do when she meets another whaler in any sort of decent weather? She has a "Gam", a thing so utterly unknown to all other ships that they never heard of the name even; and if by chance they should hear of it, they only grin at it, and repeat gamesome stuff about "spouters" and "blubber-boilers," and such like pretty exclamations. Whereas, some merchant ships crossing each other's wake in the mid-Atlantic, will oftentimes pass on without so much as a single word of recognition, mutually cutting each other on the high seas, like a brace of dandies in Broadway. As for Men-of-War, they first go through such a string of silly bowings and scrapings, such a ducking of ensigns, that there does not seem to be much right-down hearty good-will and brotherly love about it at all. As touching Slave-ships meeting, why, they are in such a prodigious hurry, they run away from each other as soon as possible.

But what is a Gam?

GAM. Noun—A social meeting of two (or more) whale-ships, generally on a cruising-ground; when, after exchanging hails, they exchange visits by boats' crews: the two captains remaining, for the time, on board of one ship, and the two chief mates on the other.

"The Gam." 1865. Gams sometimes lasted weeks at a time during which the ships, said to be mating, sailed and hunted as a team, sharing the catch. In the center of this painting is the bark *James Allen*, *built in 1844 and abandoned in the Arctic ice in 1876. It is flanked on the left by the Sea Fox. The artist, Charles S. Raleigh, much respected in the 19th century, was a master of detail. He defied anyone to observe his work and find a line in the rigging missing or out of place.*

Forbes Collection of the MIT Museum

Chapter LV

Of the Monstrous Pictures of Whales

Unascribed German lithograph, 1830, based on Cuvier's "squash." — *Kendall Collection, NBWM.*

I shall ere long paint to you as well as one can without canvas, something like the true form of the whale as he actually appears to the eye of the whaleman when in his own absolute body the whale is moored alongside the whale-ship so that he can be fairly stepped upon there. It may be worth while, therefore, previously to advert to those curious imaginary portraits of him which even down to the present day confidently challenge the faith of the landsman. It is time to set the world right in this matter, by proving such pictures of the whale all wrong.

Now, by all odds, the most ancient extant portrait anyways purporting to be the whale's, is to be found in the famous cavern-pagoda of Elephanta, in India. The Hindoo whale depicts the incarnation of Vishnu in the form of leviathan, learnedly known as the Matse Avatar.

Then, there are the Prodromus whales of the old Scotch Sibbald, and Jonah's whale, as depicted in the prints of old Bibles and the cuts of old primers. As for the book-binder's whale winding like a vine-stalk round the stock of a descending anchor, that is a very picturesque but purely fabulous creature, imitated. I nevertheless call this book-binder's fish an attempt at a whale.

But the placing of the cap-sheaf to all this blundering business was reserved for the scientific Frederick Cuvier, brother to the famous Baron. In 1836, he published a Natural History of Whales, in which he gives what he calls a picture of the sperm whale. Before showing that picture to any Nantucketer, you had best provide for your summary retreat from Nantucket. In a word, Frederick Cuvier's Sperm Whale is not a sperm whale, but a squash. Of course, he never had the benefit of a whaling voyage (such men seldom have), but whence he derived that picture, who can tell?

As for the sign-painters' whales seen in the streets hanging over the shops of oil-dealers, what shall be said of them? They are generally Richard III. whales, with dromedary humps, and very savage; breakfasting on three or four sailor tarts, that is whaleboats full of mariners: their deformities floundering in seas of blood and blue paint.

For all these reasons, then, you must needs conclude that the great leviathan is that one creature in the world which must remain unpainted to the last. And the only mode in which you can derive even a tolerable idea of his living contour, is by going a whaling yourself; but by so doing, you run no small risk of being eternally stove and sunk by him. Wherefore, it seems to me you had best not be too fastidious in your curiosity touching this leviathan.

"Greenland Whale Breaching," 1837. — Colored engraving by Robert Hamilton.

New Bedford Whaling Museum

Chapter LVI

Of the Less Erroneous Pictures of Whales, and the True Pictures of Whaling Scenes

Scoresby's right whale, 1820. Melville borrowed Scoresby's "Arctic Regions" from the New York Society Library in April, 1850 and did not return it until June, 1851. — *Source: Leyda. Engraving by W. & D. Lizars after a drawing by William Scoresby.*

I know of only four published outlines of the great Sperm Whale; Colnett's, Huggins's, Frederick Cuvier's, and Beale's. By great odds, Beale's is the best. All Beale's drawings of this whale are good, excepting the middle figure in the picture of three whales in various attitudes, capping his second chapter. His frontispiece, boats attacking sperm whales, though no doubt calculated to excite the civil scepticism of some parlor men, is admirably correct and life-like in its general effect. Some of the sperm whale drawings in J. Ross Browne are pretty correct in contour; but they are wretchedly engraved. That is not his fault though.

Of the right whale, the best outline pictures are in Scoresby; but they are drawn on too small a scale to convey a desirable impression. He has but one picture of whaling scenes, and this is a sad deficiency, because it is by such pictures only, when at all well done, that you can derive anything like a truthful idea of the living whale as seen by his living hunters.

But, taken for all in all, by far the finest, though in some details not the most correct, presentations of whales and whaling scenes to be anywhere found, are two large French engravings, well executed, and taken from paintings by one Garnery. Respectively, they represent attacks on the sperm and right whale. In the first engraving a noble sperm whale is depicted in full majesty of might, just risen beneath the boat from the profundities of the ocean, and bearing high in the air upon his back the terrific wreck of the stoven planks. The action of the whole thing is wonderfully good and true. The half-emptied line-tub floats on the whitened sea; the wooden poles of the spilled harpoons obliquely bob in it;

"Boats Attacking Whales," 1839. *This engraving is the frontispiece to Thomas Beale's manifesto, "The Natural History of the Sperm Whale." Beale was one of the first to provide an accurate description of the sperm whale's habits, biology and behavior, and his account was Melville's primary resource and companion book. During his voyages between 1830-33, Beale observed the hunt up close and ascribes disposition, personality and ingenuity to many of the animals. "No animal in creation is more monstrously ferocious," he declared. It's easy to see how Beale inspired many passages in Moby-Dick, and it's no wonder Melville kept a copy of "The Natural History" with him for many years.* — Sources: Beale, Leyda, Frank.

New Bedford Whaling Museum

the heads of the swimming crew are scattered about the whale in contrasting expressions of affright; while in the black stormy distance the ship is bearing down upon the scene. Serious fault might be found with the anatomical details of this whale, but let that pass; since, for the life of me, I could not draw so good a one.

In the second engraving, the boat is in the act of drawing alongside the barnacled flank of a large running right whale, that rolls his black weedy bulk in the sea like some mossy rock-slide from the Patagonian cliffs. His jets are erect, full, and black like soot; so that from so abounding a smoke in the chimney, you would think there must be a brave supper cooking in the great bowels below. Sea fowls are pecking at the small crabs, shell-fish, and other sea candies and maccaroni, which the right whale sometimes carries on his pestilent back. And all the while the thick-lipped leviathan is rushing through the deep, leaving tons of tumultuous white curds in his wake, and causing the slight boat to rock in the swells like a skiff caught nigh the paddle-wheels of an ocean steamer. Thus, the foreground is all raging commotion; but behind, in admirable artistic contrast, is the glassy level of a sea becalmed, the drooping unstarched sails of the powerless ship, and the inert mass of a dead whale, a conquered fortress, with the flag of capture lazily hanging from the whale-pole inserted into his spout-hole.

Who Garnery the painter is, or was, I know not. But my life for it he was either practically conversant with his subject, or else marvellously tutored by some experienced whaleman.

In addition to those fine engravings from Garnery, there are two other French engravings worthy of note, by some one who subscribes himself "H. Durand." One of them, though not precisely adapted to our present purpose, nonetheless deserves mention on other accounts. It is a quiet noon-scene among the isles of the Pacific; a French whaler anchored, inshore, in a calm, and lazily taking water on board; the loosened sails of the ship, and the long leaves of the palms in the background, both drooping together in the breezeless air. The effect is very fine, when considered with reference to its presenting the hardy fishermen under one of their few aspects of

"Peche du Cachalot" or Sperm Whale Fishery, 1834. *Colored aquatint of Ambroise Louis Garneray's (1783-1857) much heralded (and often copied and reproduced) painting. Garneray became acquainted with the sea as a mariner in the French navy during the Napoleonic Wars. He developed his craft when, as a P.O.W., he served an 8-year internment inside an old hulk moored in Portsmouth Harbor and painted portraits of his captors in return for privileges. He later earned a reputation in England as a marine painter, and when he returned to Paris in 1814, he turned professional. Known to have painted at least five whaling scenes, Garneray lived in Le Havre, a vibrant port with a whaling fleet.* — Sources: Ingalls, Frank.

New Bedford Whaling Museum

oriental repose. The other engraving is quite a different affair: the ship hove-to upon the open sea, and in the very heart of the leviathanic life, with a right whale alongside; the vessel (in the act of cutting-in) hove over to the monster as if to a quay; and a boat, hurriedly pushing off from this scene of activity, is about giving chase to whales in the distance. The harpoons and lances lie levelled for use; three oarsmen are just setting the mast in its hole; while from a sudden roll of the sea, the little craft stands half-erect out of the water, like a rearing horse. From the ship, the smoke of the torments of the boiling whale is going up like the smoke over a village of smithies; and to windward, a black cloud, rising up with earnest of squalls and rains, seems to quicken the activity of the excited seamen.

"Peche de la Baleine," or Whale Fishery, 1835. According to Melville, the French had furnished the only sketches "capable of conveying the real spirit of the whale hunt.... The English and American whale draughtsmen seem entirely content with presenting the mechanical outline of things, such as the vacant profile of the whale; which is about tantamount to sketching the profile of a pyramid." — Chapter 56. Aquatint by Ambrose Louis Garneray.

"Sperm whaling, No. 2. The Conflict," 1858. This lithograph, engraved by Albert Van Beest and J. Cole, and published by Charles Taber in New Bedford, was not well received by the discerning local public, especially whalemen, who thought it unrealistic. Taber then commissioned a new work (page 108 bottom), but hired a painter with whaling experience—Benjamin Russell—to join the team. — Sources: Kugler, Ingalls.

Spinner Collection

Kendall Collection, NBWM

"Sperm whaling, No. 1. — The Chase," 1859. *Considered one of the finest of American whaling prints, this is part of a set of two colored lithographs commissioned and published by The Charles Taber Company. Although titled No. 1, this print was produced after the two attempts at No. 2 (page 107 and below). And like No. 2, it was drawn by Albert Van Beest with R. Swain Gifford and corrected by Benjamin Russell.*

William R. Hegarty Collection

"Sperm whaling, No. 2. The Capture," 1862. *Drawn by Albert Van Beest with R. Swain Gifford and corrected by Benjamin Russell, this lithograph replaced "The Conflict." Van Beest used as his model W. J. Huggins' 1834 print, "South Sea Whale Fishery" (see p. 140) and incorporated into the background the boat being overturned in "Sperm Whaling, No. 2. The Conflict" (see p. 107). The works of Van Beest, Gifford, and Russell represent the very best of New Bedford area marine painting during the second half of the 19th century. Perhaps if Melville had known of them, he wouldn't have believed Huggins and Beale the only artists capable of accurately representing the sperm whale.*

New Bedford Whaling Museum

Stove boat, circa 1940s. Although the sperm whale's tail was the primary source of the lethal damage the whale inflicted, the image of its spiked jaw crunching whaleboats is more popularly portrayed in whalemen's art. This watercolor is one of four similar works (another is on page 201) thought to have been made aboard the *Young Phoenix*. — *Kugler:* New Bedford and Old Dartmouth.

"Then it was that monomaniac Ahab, furious with this tantalizing vicinity of his foe, which placed him all alive and helpless in the very jaws he hated; seized the long bone with his naked hands, and wildly strove to wrench it from its gripe. As now… the jaw slipped from him; the frail gunwales bent in, collapsed, and snapped, as both jaws, like an enormous shears, sliding further aft, bit the craft completely in twain, and locked themselves fast again in the sea, midway between the two floating wrecks." — *Chapter 133.*

New Bedford Whaling Museum

"Whalers in the Arctic Ice," circa 1880. Fairhaven-born artist William Bradford began his career painting ship portraits in New Bedford Harbor in the 1840s. In 1854, he persuaded Dutch immigrant Albert Van Beest to share a studio with him on the harborfront. With Van Beest, he learned to paint the great wooden ships and the sea as only the Dutch masters could. Bradford eventually developed a luminist style, painting grand landscapes emphasizing the gradual differentiation of color or light in sky, the subtle variations in the translucent greens in the seas, and the magnificent beauty of the natural world. Between 1854 and 1870, struck by the surreal beauty of the Arctic, he made nine trips there. Bradford's paintings of the Arctic usually emphasize ships dwarfed by vast icefields and towering icebergs sparkling under glowing skies. He produced and published volumes of work featuring his paintings and photographs which enjoyed considerable popularity at home and in England.

Private Collection

Chapter LVII

Of Whales in Paint; In Teeth; In Wood; In Sheet Iron; In Stone; In Mountains; In Stars

Throughout the Pacific, and also in Nantucket, and New Bedford, and Sag Harbor, you will come across lively sketches of whales and whaling-scenes, graven by the fishermen themselves on sperm whale-teeth, or ladies' busks wrought out of the right whale-bone, and other like skrimshander articles, as the whalemen call the numerous little ingenious contrivances they elaborately carve out of the rough material, in their hours of ocean leisure. Some of them have little boxes of dentistical-looking implements, specially intended for the skrimshandering business. But, in general, they toil with their jack-knives alone; and, with that almost omnipotent tool of the sailor, they will turn you out anything you please, in the way of a mariner's fancy.

Now, one of the peculiar characteristics of the savage in his domestic hours, is his wonderful patience of industry. For, with but a bit of broken sea-shell or a shark's tooth, that miraculous intricacy of wooden network has been achieved; and it has cost steady years of steady application.

As with the Hawaiian savage, so with the white sailor-savage. With the same marvellous patience, and with the same single shark's tooth, of his one poor jack-knife, he will carve you a bit of bone sculpture.

Wooden whales, or whales cut in profile out of the small dark slabs of the noble South Sea war-wood, are frequently met with in the forecastles of American whalers. Some of them are done with much accuracy.

At some old gable-roofed country houses you will see brass whales hung by the tail for knockers to the road-side door. When the porter is sleepy, the anvil-headed whale would be best. But these knocking whales are seldom remarkable as faithful essays. On the spires of some old-fashioned churches you will see sheet-iron whales placed there for weather-cocks; but they are so elevated, and besides that are to all intents and purposes so labelled with *"Hands off!"* you cannot examine them closely enough to decide upon their merit.

Kendall Collection, NBWM

New Bedford Whaling Museum

New Bedford Whaling Museum

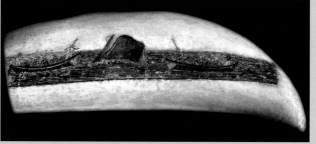
New Bedford Whaling Museum

Working with whalebone. *Top to bottom, left to right: Whalebone pie-crust crimpers; sperm whale tooth engraved by an anonymus crewman on the first voyage of the whaleship L.C. Richmond of Warren, RI, circa 1835; whalebone-trim pagoda drawer; yarn winder; sperm whale tooth.* — NBWM exhibit pieces photographed by Seth Beall.

To fill many hours of free time during long voyages, whalemen crafted and sketched whalebone into art objects of varying detail, expression and craftsmanship. Their work tells the story of their industry as well as their life and times. *"Long exile from Christendom and civilization inevitably restores a man to that condition in which God placed him, i. e., what is called savagery. I myself am a savage; owning no allegiance but to the King of the Cannibals; and ready at any moment to rebel against him. Now, one of the peculiar characteristics of the savage in his domestic hours, is his wonderful patience of industry."* — Chapter 57.

Chapter LXVIII, LX

Brit · The Line

Steering north-eastward from the Crozetts, we fell in with vast meadows of brit, the minute, yellow substance, upon which the right whale largely feeds. For leagues and leagues it undulated round us, so that we seemed to be sailing through boundless fields of ripe and golden wheat.

On the second day, numbers of right whales were seen, who, secure from the attack of a sperm whaler like the Pequod, with open jaws sluggishly swam through the brit, which, adhering to the fringing fibres of that wondrous Venetian blind in their mouths, was in that manner separated from the water that escaped at the lip.

Seen from the mast-heads, especially when they paused and were stationary for a while, their vast black forms looked more like lifeless masses of rock than anything else. And even when recognised at last, their immense magnitude renders it very hard really to believe that such bulky masses of overgrowth can possibly be instinct, in all parts, with the same sort of life that lives in a dog or a horse.

The harpoon line is carefully coiled in tubs on deck of the *John R. Manta*, 1925. — *William R. Hegarty Collection.*

CHAPTER LX

With reference to the whaling scene shortly to be described, as well as for the better understanding of all similar scenes elsewhere presented, I have here to speak of the magical, sometimes horrible whale-line.

The whale line is only two thirds of an inch in thickness. At first sight, you would not think it so strong as it really is. By experiment its one and fifty yarns will each suspend a weight of one hundred and twenty pounds; so that the whole rope will bear a strain nearly equal to three tons. In length, the common sperm whale-line measures something over two hundred fathoms. Towards the stern of the boat it is spirally coiled away in the tub, not like the worm-pipe of a still though, but so as to form one round, cheese-shaped mass of densely bedded "sheaves,"

Right whaling in the Atlantic, 1836. *"As morning mowers, who side by side slowly and seethingly advance their scythes through the long wet grass of marshy meads; even so these monsters swam, making a strange, grassy, cutting sound; and leaving behind them endless swaths of blue upon the yellow sea."* — Chapter 68. *Pen and watercolor from the journal of William W. Taylor, Second Mate, ship* South Carolina *of Dartmouth.*

Kendall Collection, NBWM

or layers of concentric spiralizations, without any hollow but the "heart," or minute vertical tube formed at the axis of the cheese. As the least tangle or kink in the coiling would, in running out, infallibly take somebody's arm, leg, or entire body off, the utmost precaution is used in stowing the line in its tub. Some harpooneers will consume almost an entire morning in this business, carrying the line high aloft and then reeving it downwards through a block towards the tub, so as in the act of coiling to free it from all possible wrinkles and twists.

The American tub, nearly three feet in diameter and of proportionate depth, makes a rather bulky freight for a craft whose planks are but one half-inch in thickness; for the bottom of the whale-boat is like critical ice, which will bear up a considerable distributed weight, but not very much of a concentrated one.

Both ends of the line are exposed; the lower end terminating in an eye-splice or loop coming up from the bottom against the side of the tub, and hanging over its edge completely disengaged from everything. This arrangement of the lower end is necessary on two accounts. First: In order to facilitate the fastening to it of an additional line from a neighboring boat, in case the stricken whale should sound so deep as to threaten to carry off the entire line originally attached to the harpoon. In these instances, the whale of course is shifted like a mug of ale, as it were, from the one boat to the other; though the first boat always hovers at hand to assist its consort. Second: This arrangement is indispensable for common safety's sake; for were the lower end of the line in any way attached to the boat, and were the whale then to run the line out to the end almost in a single, smoking minute as he sometimes does, he would not stop there, for the doomed boat would infallibly be dragged down after him into the profundity of the sea; and in that case no town-crier would ever find her again.

Before lowering the boat for the chase, the upper end of the line is taken aft from the tub, and passing round the logger-head there, is again carried forward the entire length of the boat, resting crosswise upon the loom or handle of every man's oar, so that it jogs against his wrist in rowing; and also passing between the men, as they alternately sit at the opposite gunwales, to the leaded chocks or grooves in the extreme pointed prow of the boat, where a wooden pin or skewer the size of a common quill, prevents it from slipping out. From the chocks it hangs in a slight festoon over the bows, and is then passed inside the boat again; and some ten or twenty

Havoc abaft. *With its line rapidly uncoiling and oars askew, the whaleboat is precariously balanced. "Nor can any son of mortal woman, seat himself amid those hempen intricacies, and while straining his utmost at the oar, bethink him that at any unknown instant the harpoon may be darted, and all these horrible contortions be put in play like ringed lightnings…. Yet habit—strange thing!—Gayer sallies, more merry mirth, better jokes, and brighter repartees, you never heard over your mahogany, than you will hear over the half-inch white cedar of the whale-boat, when thus hung in hangman's nooses…"* — Chapter 60. Watercolor from anonymous sketchbook from the bark Orray Taft of New Bedford, 1864-65.

Kendall Collection, NBWM

Coiling line into tubs on bark **Wanderer,** *1922.* — William H. Tripp photograph. Edwards Family Collection.

fathoms (called box-line) being coiled upon the box in the bows, it continues its way to the gunwale still a little further aft, and is then attached to the short-warp—the rope which is immediately connected with the harpoon; but previous to that connexion, the short-warp goes through sundry mystifications too tedious to detail.

Thus the whale-line folds the whole boat in its complicated coils, twisting and writhing around it in almost every direction. All the oarsmen are involved in its perilous contortions; so that to the timid eye of the landsman, they seem as Indian jugglers, with the deadliest snakes sportively festooning their limbs.

Perhaps a very little thought will now enable you to account for those repeated whaling disasters—some few of which are casually chronicled—of this man or that man being taken out of the boat by the line, and lost. For, when the line is darting out, to be seated then in the boat, is like being seated in the midst of the manifold whizzings of a steam-engine in full play, when every flying beam, and shaft, and wheel, is grazing you. It is worse; for you cannot sit motionless in the heart of these perils, because the boat is rocking like a cradle, and you are pitched one way and the other, without the slightest warning; and only by a certain self-adjusting buoyancy and simultaneousness of volition and action, can you escape being made a Mazeppa of, and run away with where the all-seeing sun himself could never pierce you out.

Again: as the profound calm which only apparently precedes and prophesies of the storm, is perhaps more awful than the storm itself; for, indeed, the calm is but the wrapper and envelope of the storm; and contains it in itself, as the seemingly harmless rifle holds the fatal powder, and the ball, and the explosion; so the graceful repose of the line, as it silently serpentines about the oarsmen before being brought into actual play—this is a thing which carries more of true terror than any other aspect of this dangerous affair.

Beetle whaleboat design, 1874. The Beetle family, known for their excellent whaleboat, and later the Beetle catboat, lived for generations at Clark's Point in New Bedford. Founder James Beetle developed "prefab" boats and mass-production methods enabling him to turn out boats quickly while maintaining the highest standards. Whaleboats could be built in one day by wrapping the planks around the skeleton, then putting in the ribs (rather than building the frame first as was usual). In 1880, a fully-equipped Beetle whaleboat cost $200. —*Source: New England Beetle Cat Boat Association web site.*

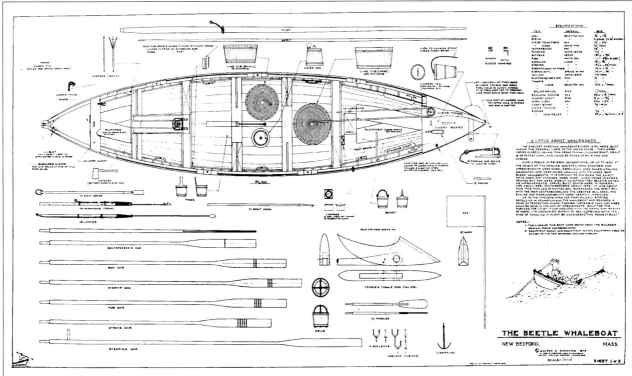

National Archives

Chapter LXI

Stubb Kills a Whale

The next day was exceedingly still and sultry, and with nothing special to engage them, the Pequod's crew could hardly resist the spell of sleep induced by such a vacant sea. For this part of the Indian Ocean through which we then were voyaging is not what whalemen call a lively ground; that is, it affords fewer glimpses of porpoises, dolphins, flying-fish, and other vivacious denizens of more stirring waters, than those off the Rio de la Plata, or the in-shore ground off Peru.

It was my turn to stand at the foremast-head; and with my shoulders leaning against the slackened royal shrouds, to and fro I idly swayed in what seemed an enchanted air.

Suddenly bubbles seemed bursting beneath my closed eyes; like vices my hands grasped the shrouds; some invisible, gracious agency preserved me; with a shock I came back to life. And lo! close under our lee, not forty fathoms off, a gigantic sperm whale lay rolling in the water like the capsized hull of a frigate, his broad, glossy back, of an Ethiopian hue, glistening in the sun's rays like a mirror.

As if struck by some enchanter's wand, the sleepy ship and every sleeper in it at once started into wakefulness; and more than a score of voices from all parts of the vessel, simultaneously with the three notes from aloft, shouted forth the accustomed cry, as the great fish slowly and regularly spouted the sparkling brine into the air.

"Clear away the boats! Luff!" cried Ahab. And obeying his own order, he dashed the helm down before the helmsman could handle the spokes.

The sudden exclamations of the crew must have alarmed the whale; and ere the boats were down, majestically turning, he swam away to the leeward, but with such a steady tranquillity, and making so few ripples as he swam, that thinking after all he might not as yet be alarmed, Ahab gave orders that not an oar should be

Sperm whale seen from the deck of schooner John R. Manta, *1925.*

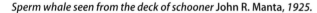
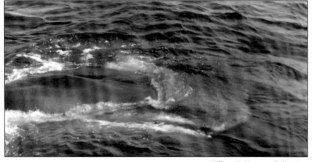

William R. Hegarty Collection

"Exploits of an American harpooner," 1861. — *From Cheevers:* Whaleman's Adventures.

used, and no man must speak but in whispers. Presently, as we thus glided in chase, the monster perpendicularly flitted his tail forty feet into the air, and then sank out of sight like a tower swallowed up.

"There go flukes!" was the cry, an announcement immediately followed by Stubb's producing his match and igniting his pipe, for now a respite was granted. After the full interval of his sounding had elapsed, the whale rose again, and being now in advance of the smoker's boat, and much nearer to it than to any of the others, Stubb counted upon the honor of the capture. It was obvious, now, that the whale had at length become aware of his pursuers. Paddles were dropped, and oars came loudly into play. And still puffing at his pipe, Stubb cheered on his crew to the assault.

Yes, a mighty change had come over the fish. All alive to his jeopardy, he was going "head out;" that part obliquely projecting from the mad yeast which he brewed.

Stubb retaining his place in the van, still encouraged his men to the onset, all the while puffing the smoke from his mouth. Like desperadoes they tugged and they strained, till the welcome cry was heard—"Stand up, Tashtego!—give it to him!" The harpoon was hurled. "Stern all!" The oarsmen backed water; the same moment something went hot and hissing along every one of their wrists. It was the magical line. As the line passed round and round the loggerhead; so also, just before reaching that point, it blisteringly passed through and through both of Stubb's hands, from which the hand-cloths, or squares of quilted canvas sometimes worn at these times, had accidentally dropped. It was like holding an enemy's sharp two-edged sword by the blade, and that enemy all the time striving to wrest it out of your clutch.

"Wet the line! wet the line!" cried Stubb to the tub oarsman (him seated by the tub) who, snatching off his hat, dashed the sea-water into it. More turns were taken, so that the line began holding its place. The boat now flew through the boiling water like a shark all fins. Stubb and Tashtego here changed places—stem for stern—a staggering business truly in that rocking commotion.

From the vibrating line extending the entire length of the upper part of the boat, and from its now being more tight than a harpstring, you would have thought the craft had two keels—one cleaving the water, the other the air—as the boat churned on through both opposing elements at once. A continual cascade played at the bows; a ceaseless whirling eddy in her wake; and, at the slightest motion from within, even but of a little finger, the vibrating, cracking craft canted over her spasmodic gunwale into the sea. Whole Atlantics and Pacifics seemed passed as they shot on their way, till at length the whale somewhat slackened his flight.

"Haul in—haul in!" cried Stubb to the bowsman! and, facing round towards the whale, all hands began pulling the boat up to him, while yet the boat was being towed on. Soon ranging up by his flank, Stubb, firmly planting his knee in the clumsy cleat, darted dart after dart into the flying fish; at the word of command, the boat alternately sterning out of the way of the whale's horrible wallow, and then ranging up for another fling.

The red tide now poured from all sides of the monster like brooks down a hill. His tormented body rolled not in brine but in blood, which bubbled and seethed for furlongs behind in their wake. The slanting sun playing upon this crimson pond in the sea, sent back its reflection into every face, so that they all glowed to each other like red men. And all the while, jet after jet of white smoke was agonizingly shot from the spiracle of the whale, and vehement puff after puff from the mouth of the excited headsman; as at every dart, hauling in upon his crooked lance (by the line attached to it), Stubb straightened it again and again, by a few rapid blows against the gunwale, then again and again sent it into the whale.

Starting from his trance into that unspeakable thing called his "flurry," the monster horribly wallowed in his blood, over-wrapped himself in impenetrable, mad, boiling spray, so that the imperilled craft, instantly dropping astern, had much ado blindly to struggle out from that phrensied twilight into the clear air of the day.

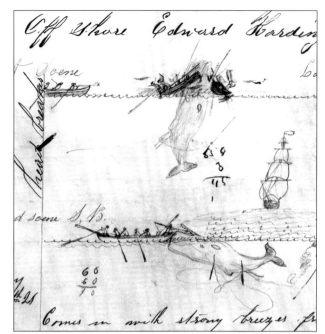

"Head breaker," 1840. Crew of the Edward Harding *wages battle.*
— *Pencil sketch from a journal aboard ship.*

And now abating in his flurry, the whale once more rolled out into view; surging from side to side; spasmodically dilating and contracting his spout-hole, with sharp, cracking, agonized respirations. At last, gush after gush of clotted red gore, as if it had been the purple lees of red wine, shot into the frighted air; and falling back again, ran dripping down his motionless flanks into the sea. His heart had burst!

"He's dead, Mr. Stubb," said Daggoo.

"Yes; both pipes smoked out!" and withdrawing his own from his mouth, Stubb scattered the dead ashes over the water; and, for a moment, stood thoughtfully eyeing the vast corpse he had made.

"Lancing a Whale," 1899. " 'Pull up—pull up!' he now cried to the bowsman, as the waning whale relaxed in his wrath. 'Pull up!—close to!' and the boat ranged along the fish's flank. When reaching far over the bow, Stubb slowly churned his long sharp lance into the fish, and kept it there, carefully churning and churning, as if cautiously seeking to feel after some gold watch that the whale might have swallowed, and which he was fearful of breaking ere he could hook it out. But that gold watch he sought was the innermost life of the fish. And now it is struck…" — *Chapter 61.*
Engraving from Bullen: Cruise of the Cachalot.

Chapter LXII

The Dart

Harpooner from the *Wanderer* ready to strike, 1923. — *Kendall Collection, NBWM.*

According to the invariable usage of the fishery, the whale-boat pushes off from the ship, with the headsman or whale-killer as temporary steersman, and the harpooneer or whale-fastener pulling the foremost oar, the one known as the harpooneer-oar. Now it needs a strong, nervous arm to strike the first iron into the fish; for often, in what is called a long dart, the heavy implement has to be flung to the distance of twenty or thirty feet. But however prolonged and exhausting the chase, the harpooneer is expected to pull his oar meanwhile to the uttermost; indeed, he is expected to set an example of superhuman activity to the rest, not only by incredible rowing, but by repeated loud and intrepid exclamations; and what it is to keep shouting at the top of one's compass, while all the other muscles are strained and half started—what that is none know but those who have tried it. In this straining, bawling state, then, with his back to the fish, all at once the exhausted harpooneer hears the exciting cry—"Stand up, and give it to him!" He now has to drop and secure his oar, turn round on his centre half way, seize his harpoon from the crotch, and with what little strength may remain, he essays to pitch it somehow into the whale. No wonder, taking the whole fleet of whalemen in a body, that out of fifty fair chances for a dart, not five are successful; no wonder that so many hapless harpooneers are madly cursed and disrated; no wonder that some of them actually burst their blood-vessels in the boat; no wonder that some sperm whalemen are absent four years with four barrels; no wonder that to many ship-owners, whaling is but a losing concern; for it is the harpooneer that makes the voyage, and if you take the breath out of his body how can you expect to find it there when most wanted!

Again, if the dart be successful, then at the second critical instant, that is, when the whale starts to run, the boat-header and harpooneer likewise start to running fore and aft, to the imminent jeopardy of themselves and every one else. It is then they change places; and the headsman, the chief officer of the little craft, takes his proper station in the bows of the boat.

Now, I care not who maintains the contrary, but all this is both foolish and unnecessary. The headsman should stay in the bows from first to last; he should both dart the harpoon and the lance, and no rowing whatever should be expected of him, except under circumstances obvious to any fisherman. I know that this would sometimes involve a slight loss of speed in the chase; but long experience in various whalemen of more than one nation has convinced me that in the vast majority of failures in the fishery, it has not by any means been so much the speed of the whale as the before described exhaustion of the harpooneer that has caused them.

Striking a sperm whale, Northwest Coast, circa 1915.

Spinner Collection

Chapter LXIII

The Crotch

"The Live Iron," circa 1938. The crotch is visible at center.
— *Sketch by George A. Gale. Kendall Collection, NBWM.*

The crotch alluded to on a previous page deserves independent mention. It is a notched stick of a peculiar form, some two feet in length, which is perpendicularly inserted into the starboard gunwale near the bow, for the purpose of furnishing a rest for the wooden extremity of the harpoon, whose other naked, barbed end slopingly projects from the prow. Thereby the weapon is instantly at hand to its hurler, who snatches it up as readily from its rest as a backwoodsman swings his rifle from the wall. It is customary to have two harpoons reposing in the crotch, respectively called the first and second irons.

But these two harpoons, each by its own cord, are both connected with the line; the object being this: to dart them both, if possible, one instantly after the other into the same whale; so that if, in the coming drag, one should draw out, the other may still retain a hold. It is a doubling of the chances. But it very often happens that owing to the instantaneous, violent, convulsive running of the whale upon receiving the first iron, it becomes impossible for the harpooneer, however lightning-like in his movements, to pitch the second iron into him. Nevertheless, as the second iron is already connected with the line, and the line is running, hence that weapon must, at all events, be anticipatingly tossed out of the boat, somehow and somewhere; else the most terrible jeopardy would involve all hands. Tumbled into the water, it accordingly is in such cases; the spare coils of box line (mentioned in a preceding chapter) making this feat, in most instances, prudently practicable. But this critical act is not always unattended with the saddest and most fatal casualties.

Furthermore: you must know that when the second iron is thrown overboard, it thenceforth becomes a dangling, sharp-edged terror, skittishly curvetting about both boat and whale, entangling the lines, or cutting them, and making a prodigious sensation in all directions. Nor, in general, is it possible to secure it again until the whale is fairly captured and a corpse.

Consider, now, how it must be in the case of four boats all engaging one unusually strong, active, and knowing whale; when owing to these qualities in him, as well as to the thousand concurring accidents of such an audacious enterprise, eight or ten loose second irons may be simultaneously dangling about him. For, of course, each boat is supplied with several harpoons to bend on to the line should the first one be ineffectually darted without recovery. All these particulars are faithfully narrated here, as they will not fail to elucidate several most important, however intricate passages, in scenes hereafter to be painted.

Ready to strike, circa 1922. Theophilo Freitas, a harpooner by trade, demonstrates his form. — *Still photograph from the film "Down to the Sea in Ships."*

New Bedford Whaling Museum

Chapter LXIV

Stubb's Supper

Stubb's whale had been killed some distance from the ship. It was a calm; so, forming a tandem of three boats, we commenced the slow business of towing the trophy to the Pequod.

Darkness came on; but three lights up and down in the Pequod's main-rigging dimly guided our way; till drawing nearer we saw Ahab dropping one of several more lanterns over the bulwarks. Vacantly eyeing the heaving whale for a moment, he issued the usual orders for securing it for the night, and then handing his lantern to a seaman, went his way into the cabin, and did not come forward again until morning.

Though, in overseeing the pursuit of this whale, Captain Ahab had evinced his customary activity, to call it so; yet now that the creature was dead, some vague dissatisfaction, or impatience, or despair, seemed working in him; as if the sight of that dead body reminded him that Moby Dick was yet to be slain; and though a thousand other whales were brought to his ship, all that would not one jot advance his grand, monomaniac object.

If moody Ahab was now all quiescence, at least so far as could be known on deck, Stubb, his second mate, flushed with conquest, betrayed an unusual but still good-natured excitement. Stubb was a high liver; he was somewhat intemperately fond of the whale as a flavorish thing to his palate.

"Launch the harpoon," 1861.
— *Engraving by W. Roberts, from Delano:* The Wanderings....

"A steak, a steak, ere I sleep! You, Daggoo! overboard you go, and cut me one from his small!"

About midnight that steak was cut and cooked; and lighted by two lanterns of sperm oil, Stubb stoutly stood up to his spermaceti supper at the capstan-head, as if that capstan were a sideboard. Nor was Stubb the only banqueter on whale's flesh that night. Mingling their mumblings with his own mastications, thousands on thousands of sharks, swarming round the dead leviathan, smackingly feasted on its fatness. The few sleepers below in their bunks were often startled by the sharp slapping of their tails against the hull, within a few inches of the sleepers' hearts. Peering over the side you could just see them (as before you heard them) wallowing in the sullen, black waters, and turning over on their backs as they scooped out huge globular pieces of the whale of the bigness of a human head.

Though amid all the smoking horror and diabolism of a sea-fight, sharks will be seen longingly gazing up to the ship's decks, like hungry dogs round a table where red meat is being carved, ready to bolt down every killed man that is tossed to them; and though sharks also are

Crew of brig Daisy towing a sperm whale, 1912. *Harpooner Antão Neves fulfills his role as boatsteerer while First Mate Almeida rests in the bow after dispatching the whale with his lance.*

New Bedford Whaling Museum

the invariable outriders of all slave ships crossing the Atlantic, systematically trotting alongside, to be handy in case a parcel is to be carried anywhere, or a dead slave to be decently buried; yet is there no conceivable time or occasion when you will find them in such countless numbers, and in gayer or more jovial spirits, than around a dead sperm whale, moored by night to a whale-ship at sea. If you have never seen that sight, then suspend your decision about the propriety of devil-worship, and the expediency of conciliating the devil.

But, as yet, Stubb heeded not the mumblings of the banquet that was going on so nigh him, no more than the sharks heeded the smacking of his own epicurean lips.

"Cook, cook!—where's that old Fleece?" he cried at length, widening his legs still further, as if to form a more secure base for his supper; and, at the same time darting his fork into the dish, as if stabbing with his lance; "cook, you cook!—sail this way, cook!"

The old black, not in any very high glee at having been previously routed from his warm hammock at a most unseasonable hour, came shambling along from his galley, for, like many old blacks, there was something the matter with his knee-pans, which he did not keep well scoured like his other pans; this old Fleece, as they called him, came shuffling and limping along, assisting his step with his tongs, which, after a clumsy fashion, were made of straightened iron hoops; this old Ebony floundered along, and in obedience to the word of command, came to a dead stop on the opposite side of Stubb's sideboard; when, with both hands folded before him, and resting on his two-legged cane, he bowed his arched back still further over, at the same time sideways inclining his head, so as to bring his best ear into play.

"Cook," said Stubb, "Don't I always say that to be good, a whale-steak must be tough? There are those sharks now over the side, don't you see, they prefer it tough and rare? Cook, go and talk to 'em; tell 'em they are welcome to help themselves civilly, and in moderation, but they must keep quiet. Here, take this lantern," snatching one from his sideboard; "now then, go and preach to 'em!"

Sullenly taking the offered lantern, old Fleece limped across the deck to the bulwarks; and then, with one hand dropping his light low over the sea, so as to get a good view of his congregation, with the other hand he solemnly flourished his tongs, and leaning far over the side in a mumbling voice began addressing the sharks, while Stubb, softly crawling behind, overheard all that was said.

"Fellow-critters: I'se ordered here to say dat you must stop dat dam noise dare. you hear? stop dat dam smackin' ob de lip! massa Stubb say dat you can fill your dam bellies up to de hatchings, but by Gor! you must stop dat dam racket!"

Crew of the John R. Manta *line up for hardtack, circa 1925.* "We had crackers and salted beef with coffee for breakfast…soup for lunch and supper. We also had breast of whale sometimes, with a little salt. Sometimes, we fished over the side for food. After several months, our cook began to lose his mind. He stopped cooking—didn't know what he was doing. So another man took over. But when he cooked the soup, it was like water. So one day, we went on strike. We had to have our strength, you know." — *Source: Joe Ramos interview,* Spinner II. *Photograph by Capt. Henry Mandley. New Bedford Whaling Museum.*

"Cook," here interposed Stubb, accompanying the word with a sudden slap on the shoulder,— "Cook! why, damn your eyes, you mustn't swear that way when you're preaching. That's no way to convert sinners, Cook!"

"Who dat? Den preach to him yourself," sullenly turning to go.

"No, Cook; go on, go on."

"Well, den, Belubed fellow-critters:—"

"Right!" exclaimed Stubb, approvingly, "coax 'em to it; try that," and Fleece continued.

"Your woraciousness, fellow-critters, I don't blame ye so much for; dat is natur, and can't be helped; but to gobern dat wicked natur, dat is de pint. You is sharks, sartin; but if you gobern de shark in you, why den you be angel; for all angel is not'ing more dan de shark well goberned. Now, look here, bred'ren, just try wonst to be cibil, a helping yourselbs from dat whale. Don't be tearin' de blubber out your neighbour's mout, I say. Is not one shark dood right as toder to dat whale? And, by Gor, none on you has de right to dat whale; dat whale belong to some one else. I know some o' you has berry brig mout, brigger dan oders; but den de brig mouts sometimes has de small bellies; so dat de brigness ob de mout is not to swallar wid, but to bite off de blubber for de small fry ob sharks, dat can't get into de scrouge to help demselves."

"Well done, old Fleece!" cried Stubb, "that's Christianity; go on."

"No use goin' on; de dam willains will keep a scrougin' and slappin' each oder, Massa Stubb; dey don't hear one word; no use a-preachin' to such dam g'uttons as you call 'em, till dare bellies is full, and dare bellies is bottomless; and when dey do get em full, dey wont hear you den; for den dey sink in de sea, go fast to sleep on de coral, and can't hear not'ing at all, no more, for eber and eber."

"Upon my soul, I am about of the same opinion; so give the benediction, Fleece, and I'll away to my supper."

Upon this, Fleece, holding both hands over the fishy mob, raised his shrill voice, and cried—

"Cussed fellow-critters! Kick up de damndest row as ever you can; fill your dam' bellies 'till dey bust—and den die."

"Now, cook," said Stubb, resuming his supper at the capstan; "Stand just where you stood before, there, over against me, and pay particular attention." "All 'dention," said Fleece, again stooping over upon his tongs in the desired position.

"Well," said Stubb, helping himself freely meanwhile; "I shall now go back to the subject of this steak. In the first place, how old are you, cook?"

"What dat do wid de 'teak," said the old black, testily.

"Silence! How old are you, cook?"

"'Bout ninety, dey say," he gloomily muttered.

"And have you lived in this world hard upon one hundred years, cook, and don't know yet how to cook a whale-steak?" rapidly bolting another mouthful at the last word, so that that morsel seemed a continuation of the question. "Where were you born, cook?"

"'Hind de hatchway, in ferry-boat, goin' ober de Roanoke."

"Born in a ferry-boat! That's queer, too. But I want to know what country you were born in, cook?"

"Didn't I say de Roanoke country?" he cried, sharply.

"No, you didn't, cook; but I'll tell you what I'm coming to, cook. You must go home and be born over again; you don't know how to cook a whale-steak yet."

"Bress my soul, if I cook noder one," he growled, angrily, turning round to depart.

"Come back, cook;—here, hand me those tongs;—now take that bit of steak there, and tell me if you think that steak cooked as it should be? Take it, I say"—holding the tongs towards him— "take it, and taste it."

Faintly smacking his withered lips over it for a moment, the old negro muttered, "Best cooked 'teak I eber taste; joosy, berry joosy."

"Cook," said Stubb, squaring himself once more; "do you belong to the church?"

"Passed one once in Cape-Down," said the old man sullenly.

"And you have once in your life passed a holy church in Cape-Town, where you doubtless overheard a holy parson addressing his hearers as his beloved fellow-creatures, have you, cook! And yet you come here, and tell me such a dreadful lie as you did just now, eh?" said Stubb. "Where do you expect to go to, cook?" "Go to bed berry soon," he mumbled, half-turning as he spoke.

"Avast! heave to! I mean when you die, cook. It's an awful question. Now what's your answer?"

"When dis old brack man dies," said the negro slowly, changing his whole air and demeanor, "he hisself won't go nowhere; but some bressed angel will come and fetch him."

"Fetch him? How? In a coach and four, as they fetched Elijah? And fetch him where?"

"Up dere," said Fleece, holding his tongs straight over his head, and keeping it there very solemnly.

"So, then, you expect to go up into our main-top, do you, cook, when you are dead? But don't you know the higher you climb, the colder it gets? Main-top, eh?"

"Didn't say dat t'all," said Fleece, again in the sulks.

"You said up there, didn't you, and now look yourself, and see where your tongs are pointing. But, perhaps you expect to get into heaven by crawling through the lubber's hole, cook; but no, no, cook, you don't get there, except you go the regular way, round by the rigging. It's a ticklish business, but must be done, or else it's no go. But none of us are in heaven yet. Drop your tongs, cook, and hear my orders. Do ye hear? Hold your hat in one hand, and clap t'other a'top of your heart, when I'm giving my orders, cook. What! that your heart, there?—that's your gizzard! Aloft! aloft!—that's it—now you have it. Hold it there now, and pay attention."

"All 'dention," said the old black, with both hands placed as desired, vainly wriggling his grizzled head, as if to get both ears in front at one and the same time.

"Well then, cook; you see this whale-steak of yours was so very bad, that I have put it out of sight as soon as possible; you see that, don't you? Well, for the future, when you cook another whale-steak for my private table here, the capstan, I'll tell you what to do so as not to spoil it by overdoing. Hold the steak in one hand, and show a live coal to it with the other; that done, dish it; d'ye hear? And now to-morrow, cook, when we are cutting in the fish, be sure you stand by to get the tips of his fins; have them put in pickle. As for the ends of the flukes, have them soused, cook. There, now ye may go." But Fleece had hardly got three paces off, when he was recalled.

"Cook, give me cutlets for supper to-morrow night in the mid-watch. D'ye hear? away you sail, then.—Halloa! stop! make a bow before you go.—Avast heaving again! Whale-balls for breakfast—don't forget."

"Wish, by gor! whale eat him, 'stead of him eat whale. I'm bressed if he ain't more of shark dan Massa Shark hisself," muttered the old man, limping away; with which sage ejaculation he went to his hammock.

Chapter LXV

The 'Shark Massacre'

Cutting-in aboard the *Chelsea*, circa 1873. — *Davis:* Nimrod of the Sea.

When in the Southern Fishery, a captured sperm whale, after long and weary toil, is brought alongside late at night, it is not, as a general thing at least, customary to proceed at once to the business of cutting him in. For that business is an exceedingly laborious one; is not very soon completed; and requires all hands to set about it. Therefore, the common usage is to take in all sail; lash the helm a'lee; and then send every one below to his hammock till daylight, with the reservation that, until that time, anchor-watches shall be kept; that is, two and two for an hour, each couple, the crew in rotation shall mount the deck to see that all goes well.

But sometimes, especially upon the Line in the Pacific, this plan will not answer at all; because such incalculable hosts of sharks gather round the moored carcase, that were he left so for six hours, say, on a stretch, little more than the skeleton would be visible by morning. In most other parts of the ocean, however, where these fish do not so largely abound, their wondrous voracity can be at times considerably diminished, by vigorously stirring them up with sharp whaling-spades, a procedure notwithstanding, which, in some instances, only seems to tickle them into still greater activity.

Nevertheless, upon Stubb setting the anchor-watch after his supper was concluded; and when, accordingly, Queequeg and a forecastle seaman came on deck, no small excitement was created among the sharks; for immediately suspending the cutting stages over the side, and lowering three lanterns, so that they cast long gleams of light over the turbid sea, these two mariners, darting their long whaling-spades, kept up an incessant murdering of the sharks,[1] by striking the keen steel deep into their skulls. Killed and hoisted on deck for the sake of his skin, one of these sharks almost took poor Queequeg's hand off, when he tried to shut down the dead lid of his murderous jaw.

---

[1] The whaling-spade used for cutting-in is made of the very best steel; is about the bigness of a man's spread hand; and in general shape, corresponds to the garden implement after which it is named; only its sides are perfectly flat, and its upper end considerably narrower than the lower. This weapon is always kept as sharp as possible; and when being used is occasionally honed, just like a razor. In its socket, a stiff pole, from twenty to thirty feet long, is inserted for a handle.

Cutting-in a sperm whale aboard the John R. Manta, 1925. *From the cutting-in staging, two of the ship's mates cut into the whale. Only officers could be positioned above the carcass to begin the process.*

William R. Hegarty Collection

Chapter LXVII, LXVIII, LXIX

Cutting In · The Blanket · The Funeral

It was a Saturday night, and such a Sabbath as followed! The ivory Pequod was turned into what seemed a shamble; every sailor a butcher. You would have thought we were offering up ten thousand red oxen to the sea gods.

In the first place, the enormous cutting tackles, among other ponderous things comprising a cluster of blocks generally painted green, and which no single man can possibly lift—this vast bunch of grapes was swayed up to the main-top and firmly lashed to the lower mast-head, the strongest point anywhere above a ship's deck. The end of the hawser-like rope winding through these intricacies, was then conducted to the windlass, and the huge lower block of the tackles was swung over the whale; to this block the great blubber hook, weighing some one hundred pounds, was attached. And now suspended in stages over the side, Starbuck and Stubb, the mates, armed with their long spades, began cutting a hole in the body for the insertion of the hook just above the nearest of the two side-fins. This done, a broad, semicircular line is cut round the hole, the hook is inserted, and the main body of the crew striking up a wild chorus, now commence heaving in one dense crowd at the windlass. When instantly, the entire ship careens over on her side; every bolt in her starts like the nail-heads of an old house in frosty weather; she trembles, quivers, and nods her frighted mast-heads to the sky. More and more she leans over to the whale, while every gasping heave of the windlass is answered by a helping heave from the billows; till at last, a swift, startling snap is heard; with a great swash the ship rolls upwards and backwards from the whale,

Hoisting blubber, 1873.
— *Davis:* Nimrod of the Sea.

Boarding the blanket on the John R. Manta, *1925.* *Typically, two systems of blocks and tackle, called "cutting-in falls," were used for raising the blanket pieces. When the blubber was hoisted as high as it could go, a boarding knife—a sword-like blade mounted on a wood handle—was used to cut a new hole in the lower end of the blanket piece into which another hook was inserted. A cut above the second hook released the blanket from the whale, and the blanket piece was swung inboard and dropped to the deck. This process was known as boarding the blanket. Third mate Mr. Duarte, seen below, uses a boarding knife to cut a new hole into the blubber.*

Stripping blubber on the Charles W. Morgan, *circa 1912.* *A hole is cut into the blubber near the head, and the blubber hook, suspended from the mast, is inserted. The blubber is hoisted up while the men keep cutting and the carcass rolls over in the water as the blanket piece is hauled up.*

Kendall Collection, NBWM

William R. Hegarty Collection

and the triumphant tackle rises into sight dragging after it the disengaged semicircular end of the first strip of blubber. Now as the blubber envelopes the whale precisely as the rind does an orange, so is it stripped off from the body precisely as an orange is sometimes stripped by spiralizing it. For the strain constantly kept up by the windlass continually keeps the whale rolling over and over in the water, and as the blubber in one strip uniformly peels off along the line called the "scarf," simultaneously cut by the spades of Starbuck and Stubb, the mates; and just as fast as it is thus peeled off, and indeed by that very act itself, it is all the time being hoisted higher and higher aloft till its upper end grazes the main-top; the men at the windlass then cease heaving, and for a moment or two the prodigious blood-dripping mass sways to and fro as if let down from the sky, and every one present must take good heed to dodge it when it swings, else it may box his ears and pitch him headlong overboard.

One of the attending harpooneers now advances with a long, keen weapon called a boarding-sword, and watching his chance he dexterously slices out a considerable hole in the lower part of the swaying mass. Into this hole, the end of the second alternating great tackle is then hooked so as to retain a hold upon the blubber, in order to prepare for what follows. Whereupon, this accomplished swordsman, warning all hands to stand off, once more makes a scientific dash at the mass, and with a few sidelong, desperate, lunging slicings, severs it completely in twain; so that while the short lower part is still fast, the long upper strip, called a blanket-piece, swings clear, and is all ready for lowering. The heavers forward now resume their song, and while the one tackle is peeling and hoisting a second strip from the whale, the other is slowly slackened away, and down goes the first strip through the main hatchway right beneath, into an unfurnished parlor called the blubber-room. Into this twilight apartment sundry nimble hands keep coiling away the long blanket-piece as if it were a great live mass of plaited serpents. And thus the work proceeds; the two tackles hoisting and lowering simultaneously; both whale and windlass heaving, the heavers singing, the blubber-room gentlemen coiling, the mates scarfing, the ship straining, and all hands swearing occasionally, by way of assuaging the general friction.

Chapter LXVIII

The question is, what and where is the skin of the whale? Already you know what his blubber is. That blubber is something of the consistence of firm, close-grained beef, but tougher, more elastic and compact, and ranges from eight or ten to twelve and fifteen inches in thickness.

Assuming the blubber to be the skin of the whale; then, when this skin, as in the case of a very large Sperm Whale, will yield the bulk of one hundred barrels of oil; and, when it is considered that, in quantity, or rather weight, that oil, in its expressed state, is only three fourths, and not the entire substance of the coat; some idea may hence be had of the enormousness of that animated mass, a mere part of whose mere integument yields such a lake of liquid as that. Reckoning ten barrels to the ton, you have ten tons for the net weight of only three quarters of the stuff of the whale's skin.

A word or two more concerning this matter of the skin or blubber of the whale. It has already been said, that it is stript from him in long pieces, called blanket-pieces. Like most sea-terms, this one is very happy and significant. For the whale is indeed wrapt up in his blubber as in a real blanket or counterpane; or, still better, an Indian poncho slipt over his head, and skirting his extremity. It is by reason of this cosy blanketing of his body, that the whale is enabled to keep himself comfortable in all weathers, in all seas, times, and tides.

Chapter LXVII

"Haul in the chains! Let the carcase go astern!" The vast tackles have now done their duty. The peeled white body of the beheaded whale flashes like a marble sepulchre; though changed in hue, it has not perceptibly lost anything in bulk. It is still colossal. Slowly it floats more and more away, the water round it torn and splashed by the insatiate sharks, and the air above vexed with rapacious flights of screaming fowls, whose beaks are like so many insulting poniards in the whale. The vast white headless phantom floats further and further from the ship, and every rod that it so floats, what seem square roods of sharks and cubic roods of fowls, augment the murderous din.

Thus, while in life the great whale's body may have been a real terror to his foes, in his death his ghost becomes a powerless panic to a world.

"Carcass afloat," 1873. — Davis: Nimrod of the Sea.

Chapter LXX

The Sphynx

It should not have been omitted that previous to completely stripping the body of the leviathan, he was beheaded. Now, the beheading of the sperm whale is a scientific anatomical feat, upon which experienced whale surgeons pride themselves: and not without reason.

Consider that the whale has nothing that can properly be called a neck; on the contrary, where his head and body seem to join, there, in that very place, is the thickest part of him. Remember, also, that the surgeon must operate from above, some eight or ten feet intervening between him and his subject, and that subject almost hidden in a discolored, rolling, and oftentimes tumultuous and bursting sea. Bear in mind, too, that under these untoward circumstances he has to cut many feet deep in the flesh; and in that subterranean manner, without so much as getting one single peep into the ever-contracting gash thus made, he must skilfully steer clear of all adjacent, interdicted parts, and exactly divide the spine at a critical point hard by its insertion into the skull. Do you not marvel, then, at Stubb's boast, that he demanded but ten minutes to behead a Sperm Whale?

The Pequod's whale being decapitated and the body stripped, the head was hoisted against the ship's side—about half way out of the sea, so that it might yet in great part be buoyed up by its native element. And there with the strained craft steeply leaning over to it, by reason of the enormous downward drag from the lower mast-head, and every yard-arm on that side projecting like a crane over the waves; there, that blood-dripping head hung to the Pequod's waist like the giant Holofernes's from the girdle of Judith.

Two small heads on deck of the *John R. Manta, 1925.* — *William R. Hegarty Collection.*

When this last task was accomplished it was noon, and the seamen went below to their dinner. Silence reigned over the before tumultuous but now deserted deck.

A short space elapsed, and up into this noiselessness came Ahab alone from his cabin. Taking a few turns on the quarter-deck, he paused to gaze over the side, then slowly getting into the main-chains he took Stubb's long spade—still remaining there after the whale's decapitation—and striking it into the lower part of the half-suspended mass, placed its other end crutch-wise under one arm, and so stood leaning over with eyes attentively fixed on this head.

It was a black and hooded head; and hanging there in the midst of so intense a calm, it seemed the Sphynx's in the desert. "Speak, thou vast and venerable head," muttered Ahab, "which, though ungarnished with a beard, yet here and there lookest hoary with mosses; speak, mighty head, and tell us the secret thing that is in thee."

Hauling the case on the Sunbeam, 1904. *The blubber is cut away (left) and an incision made into the head with the "head spade," a heavy-gauged spade sharpened on three edges and wielded with force to facilitate the decapitation. After the head is severed, a chain is threaded through the forehead with a "head needle," and the case is hauled on deck. With a large whale such as the 85-barrel sperm at right, the head is sliced and the "junk" brought up separately. The junk is the wedge-shaped, lower half of the whale's forehead made of meat, oil and spermacetti.* — *Photographs by Clifford W. Ashley.*

New Bedford Whaling Museum

Kendall Collection, NBWM

Kendall Collection, NBWM

Chapter LXXII

The Monkey Rope

It was mentioned that upon first breaking ground in the whale's back, the blubber-hook was inserted into the original hole there cut by the spades of the mates. But in very many cases, circumstances require that the harpooneer shall remain on the whale till the whole flensing or stripping operation is concluded. The whale be it observed, lies almost entirely submerged, excepting the immediate parts operated upon. So down there, some ten feet below the level of the deck, the poor harpooneer flounders about, half on the whale and half in the water, as the vast mass revolves like a tread-mill beneath him.

You have seen Italian organ-boys holding a dancing-ape by a long cord. Just so, from the ship's steep side, did I hold Queequeg down there in the sea, by what is technically called in the fishery a monkey-rope, attached to a strong strip of canvas belted round his waist.

It was a humorously perilous business for both of us. For, before we proceed further, it must be said that the monkey-rope was fast at both ends; fast to Queequeg's broad canvas belt, and fast to my narrow leather one. But handle Queequeg's monkey-rope heedfully as I would, sometimes he jerked it so, that I came very near sliding overboard. Nor could I possibly forget that, do what I would, I only had the management of one end of it.

"Cutting in the Whale," 1905.
— *Oil painting by Clifford W. Ashley.
New Bedford Free Public Library.*

I would often jerk poor Queequeg from between the whale and the ship—where he would occasionally fall, from the incessant rolling and swaying of both. But this was not the only jamming jeopardy he was exposed to. Unappalled by the massacre made upon them during the night, the sharks now freshly and more keenly allured by the before pent blood which began to flow from the carcase—the rabid creatures swarmed round it like bees in a beehive.

Accordingly, besides the monkey-rope, with which I now and then jerked the poor fellow from too close a vicinity to the maw of what seemed a peculiarly ferocious shark—he was provided with still another protection. Suspended over the side in one of the stages, Tashtego and Daggoo continually flourished over his head a couple of keen whale-spades, wherewith they slaughtered as many sharks as they could reach. But poor Queequeg, I suppose, straining and gasping there with that great iron hook—poor Queequeg, I suppose, only prayed to his Yojo, and gave up his life into the hands of his gods.

Cutting in on the Charles W. Morgan, *circa 1907.* *While the blanket piece is being hoisted on deck, the mate on the planking begins cutting into the head casing. Captain James Earle, who took this photograph, is noted for making the largest catch of ambergris ever recorded, a feat accomplished while commanding the ship* Splendid *of Dunedin, New Zealand in 1883.*

Spinner Collection

Chapter LXXIII

Stubb and Flask Kill a Right Whale; and Then Have a Talk Over Him

Hunting the right whale. — *Davis: Nimrod of the Sea.*

Now, during the past night and forenoon, the Pequod had gradually drifted into a sea, which, by its occasional patches of yellow brit, gave unusual tokens of the vicinity of right whales. And though the Pequod was not commissioned to cruise for them at all; yet now that a sperm whale had been brought alongside and beheaded, to the surprise of all, the announcement was made that a right whale should be captured that day, if opportunity offered.

Nor was this long wanting. Tall spouts were seen to leeward; and two boats, Stubb's and Flask's, were detached in pursuit. Pulling further and further away, they at last became almost invisible to the men at the mast-head. But suddenly in the distance, they saw a great heap of tumultuous white water, and soon after news came from aloft that one or both the boats must be fast. An interval passed and the boats were in plain sight, in the act of being dragged right towards the ship by the towing whale. So close did the monster come to the hull, that at first it seemed as if he meant it malice; but suddenly going down in a maelstrom, within three rods of the planks, he wholly disappeared from view, as if diving under the keel. "Cut, cut!" was the cry from the ship to the boats, which, for one instant, seemed on the point of being brought with a deadly dash against the vessel's side. But having plenty of line yet in the tubs, and the whale not sounding very rapidly, they paid out abundance of rope, and at the same time pulled with all their might so as to get ahead of the ship. For a few minutes the struggle was intensely critical; for while they still slacked out the tightened line in one direction, and still plied their oars in another, the contending strain threatened to take them under. But it was only a few feet advance they sought to gain. And they stuck to it till they did gain it; when instantly, a swift tremor was felt running like lightning along the keel, as the strained line, scraping beneath the ship, suddenly rose to view under her bows, snapping and quivering; and so flinging off its drippings, that the drops fell like bits of broken glass on the water, while the whale beyond also rose to sight, and once more the boats were free to fly. But the fagged whale abated his speed, and blindly altering his course, went round the stern of the ship towing the two boats after him, so that they performed a complete circuit.

Meantime, they hauled more and more upon their lines, till close flanking him on both sides, Stubb answered Flask with lance for lance; and thus round and round the Pequod the battle went, while the multitudes of sharks that had before swum round the sperm whale's body, rushed to the fresh blood that was spilled, thirstily drinking at every new gash.

At last his spout grew thick, and with a frightful roll and vomit, he turned upon his back a corpse.

While the two headsmen were engaged in making fast cords to his flukes, and in other ways getting the mass in readiness for towing, some conversation ensued between them.

"I wonder what the old man wants with this lump of foul lard," said Stubb, not without some disgust at the thought of having to do with so ignoble a leviathan.

"Wants with it?" said Flask, coiling some spare line in the boat's bow, "did you never hear that the ship which but once has a sperm whale's head hoisted on her starboard side, and at the same time a right whale's on the larboard; did you never hear, Stubb, that that ship can never afterwards capsize?"

"Why not?"

"I don't know, but I heard that gamboge ghost of a Fedallah saying so, and he seems to know all about ships' charms. But I sometimes think he'll charm the ship to no good at last. I don't half like that chap, Stubb. Did you ever notice how that tusk of his is a sort of carved into a snake's head, Stubb?"

"Sink him! I never look at him at all; but if ever I get a chance of a dark night, and he standing hard by the bulwarks, and no one by; look down there, Flask"—pointing into the sea with a peculiar motion of both hands—"Aye, will I! Flask, I take that Fedallah to be the devil in disguise. Do you believe that cock and bull story about his having been stowed away on board ship? He's the devil, I say. The reason why you don't see his tail, is because he tucks it up out of sight; he carries it coiled away in his pocket, I guess. Blast him! now that I think of it, he's always wanting oakum to stuff into the toes of his boots."

"He sleeps in his boots, don't he? He hasn't got any hammock; but I've seen him lay nights in a coil of rigging."

"No doubt, and it's because of his cursed tail; he coils it down, do ye see, in the eye of the rigging."

"What's the old man have so much to do with him for?"

"Striking up a swap or a bargain, I suppose."

"Bargain?—about what?"

"Why, do ye see, the old man is hard bent after that white whale, and the devil there is trying to come round him, and get him to swap away his silver watch, or his soul, or something of that sort, and then he'll surrender Moby Dick."

"Pooh! Stubb, you are skylarking; how can Fedallah do that?"

"I don't know, Flask, but the devil is a curious chap, and a wicked one, I tell ye. But look sharp—aint you all ready there? Well, then, pull ahead, and let's get the whale alongside."

"But now, tell me, Stubb, do you suppose that that devil you was speaking of just now, was the same you say is now on board the Pequod?"

"Am I the same man that helped kill this whale? Doesn't the devil live for ever; who ever heard that the devil was dead? Did you ever see any parson a wearing mourning for the devil? And if the devil has a latch-key to get into the admiral's cabin, don't you suppose he can crawl into a port-hole? Tell me that, Mr. Flask?"

"How old do you suppose Fedallah is, Stubb?"

"Do you see that mainmast there?" pointing to the ship; "well, that's the figure one; now take all the hoops in the Pequod's hold, and string 'em along in a row with that mast, for oughts, do you see; well, that wouldn't begin to be Fedallah's age."

"But see here, Stubb, I thought you a little boasted just now, that you meant to give Fedallah a sea-toss, if you got a good chance. Now, if he's so old as all those hoops of yours come to, and if he is going to live for ever, what good will it do to pitch him overboard—tell me that?"

"Give him a good ducking, anyhow."

"But he'd crawl back."

"Duck him again; and keep ducking him."

"Suppose he should take it into his head to duck you, though—yes, and drown you—what then?"

"I should like to see him try it; I'd give him such a pair of black eyes that he wouldn't dare to show his face in the admiral's cabin again for a long while, let alone down in the orlop there, where he lives, and hereabouts on the upper decks where he sneaks so much. Damn the devil, Flask; do you suppose I'm afraid of the devil? Who's afraid of him, except the old governor who daresn't catch him and put him in double-darbies, as he deserves, but lets him go about kidnapping people? There's a governor!"

"Do you suppose Fedallah wants to kidnap Captain Ahab?"

"Do I suppose it? You'll know it before long, Flask. But I am going now to keep a sharp look-out on him; and if I see anything very suspicious going on, I'll just take him by the nape of his neck, and say—Look here, Beelzebub, you don't do it; and if he makes any fuss, by the Lord I'll make a grab into his pocket for his tail, take it to the capstan, and give him such a wrenching and heaving, that his tail will come short off at the stump—do you see; and then, I rather guess when he finds himself docked in that queer fashion, he'll sneak off without the poor satisfaction of feeling his tail between his legs."

"And what will you do with the tail, Stubb?"

"Do with it? Sell it for an ox whip when we get home;—what else?"

"Now, do you mean what you say, and have been saying all along, Stubb?"

"Mean or not mean, here we are at the ship."

The boats were here hailed, to tow the whale on the larboard side, where fluke chains and other necessaries were already prepared for securing him.

"Didn't I tell you so?" said Flask; "yes, you'll soon see this right whale's head hoisted up opposite that parmacetti's."

In good time, Flask's saying proved true. As before, the Pequod steeply leaned over towards the sperm whale's head, now, by the counterpoise of both heads, she regained her even keel; though sorely strained, you may well believe. So, when on one side you hoist in Locke's head, you go over that way; but now, on the other side, hoist in Kant's and you come back again; but in very poor plight. Thus, some minds for ever keep trimming boat. Oh, ye foolish! throw all these thunder-heads overboard, and then you will float light and right.

In disposing of the body of a right whale, when brought alongside the ship, the same preliminary proceedings commonly take place as in the case of a sperm whale; only, in the latter instance, the head is cut off whole, but in the former the lips and tongue are separately removed and hoisted on deck, with all the well known black bone attached to what is called the crown-piece. But nothing like this, in the present case, had been done. The carcases of both whales had dropped astern; and the head-laden ship not a little resembled a mule carrying a pair of overburdening panniers.

Meantime, Fedallah was calmly eyeing the right whale's head, and ever and anon glancing from the deep wrinkles there to the lines in his own hand. And Ahab chanced so to stand, that the Parsee occupied his shadow; while, if the Parsee's shadow was there at all it seemed only to blend with, and lengthen Ahab's. As the crew toiled on, Laplandish speculations were bandied among them, concerning all these passing things.

Chapter LXXIV

The Sperm Whale's Head— Contrasted View

Here, now, are two great whales, laying their heads together. Both are massive enough in all conscience; but there is a certain mathematical symmetry in the sperm whale's which the right whale's sadly lacks. Of the grand order of folio leviathans, the sperm whale and the right whale are by far the most noteworthy. To the Nantucketer, they present the two extremes of all the known varieties of the whale. The external difference between them is mainly observable in the their heads.

Let us note what is least dissimilar in these heads— namely, the two most important organs, the eye and the ear. Far back on the side of the head, and low down, near the angle of either whale's jaw, if you narrowly search, you will at last see a lashless eye, which you would fancy to be a young colt's eye; so out of all proportion is it to the magnitude of the head.

Now, from this peculiar sideway position of the whale's eyes, it is plain that he can never see an object which is exactly ahead. Moreover, the peculiar position of the whale's eyes, effectually divided as they are by many cubic feet of solid head, must wholly separate the impressions which each independent organ imparts. The whale, therefore, must see one distinct picture on this side, and another distinct picture on that side; while all between must be profound darkness and nothingness to him. This peculiarity of the whale's eyes is a thing always to be borne in mind in the fishery; and to be remembered by the reader in some subsequent scenes.

How is it, then, with the whale? True, both his eyes, in themselves, must simultaneously act; but is his brain so much more comprehensive, combining, and subtle than man's, that he can at the same moment of time attentively examine two distinct prospects, one on one side of him, and the other in an exactly opposite direction? If he can, then is it as marvellous a thing in him, as if a man were able simultaneously to go through the demonstrations of two distinct problems in Euclid.

But the ear of the whale is full as curious as the eye. If you are an entire stranger to their race, you might hunt over these two heads for hours, and never discover that organ. Is it not curious, that so vast a being as the whale should see the world through so small an eye, and hear the thunder through an ear which is smaller than a hare's?

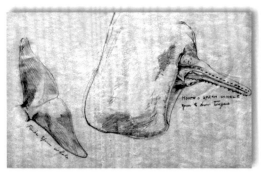

Flukes and open mouth (to show tongue) of a sperm whale, circa 1900.
— *Pencil sketch by Clement Nye Swift. Kendall Collection, NBWM.*

Let us now have a peep down the mouth; and were it not that the body is now completely separated from it, with a lantern we might descend into the great Kentucky Mammoth Cave of his stomach. But let us hold on here by this tooth, and look about us where we are. What a really beautiful and chaste- looking mouth! from floor to ceiling, lined, or rather papered with a glistening white membrane, glossy as bridal satins.

But come out now, and look at this portentous lower jaw, which seems like the long narrow lid of an immense snuff-box. If you pry it up, so as to get it overhead, and expose its rows of teeth, it seems a terrific portcullis. But far more terrible is it to behold, when fathoms down in the sea, you see some sulky whale, floating there suspended, with his prodigious jaw, some fifteen feet long, hanging straight down at right-angles with his body, for all the world like a ship's jib-boom. This whale is not dead; he is only dispirited; out of sorts, perhaps; hypochondriac; and so supine, that the hinges of his jaw have relaxed, leaving him there in that ungainly sort of plight.

In most cases this lower jaw—being easily unhinged by a practised artist—is disengaged and hoisted on deck for the purpose of extracting the ivory teeth, and furnishing a supply of that hard white whalebone with which the fishermen fashion all sorts of curious articles, including canes, umbrella-stocks, and handles to riding-whips.

With a long, weary hoist the jaw is dragged on board, as if it were an anchor; and when the proper time comes—Queequeg, Daggoo, and Tashtego, being all accomplished dentists, are set to drawing teeth. With a keen cutting-spade, Queequeg lances the gums; then the jaw is lashed down to ringbolts, and a tackle being rigged from aloft, they drag out these teeth, as Michigan oxen drag stumps of old oaks out of wild wood-lands. There are generally forty-two teeth in all; in old whales, much worn down, but undecayed; nor filled after our artificial fashion. The jaw is afterwards sawn into slabs, and piled away like joists for building houses.

Chapter LXXV

The Right Whale's Head—Contrasted View

Right whale head-on with lips closed.
— *Davis:* Nimrod of the Sea.

Crossing the deck, let us now have a good long look at the right whale's head.

As in general shape the noble sperm whale's head may be compared to a Roman war-chariot, the right whale's head bears a rather inelegant resemblance to a gigantic galliot-toed shoe.

But as you come nearer to this great head it begins to assume different aspects, according to your point of view. If you fix your eye upon this strange, crested, comb-like incrustation on the top of the mass—this green, barnacled thing, which the Greenlanders call the "crown," and the Southern fishers the "bonnet" of the right whale, you would take the head for the trunk of some huge oak, with a bird's nest in its crotch. But if this whale be a king, he is a very sulky looking fellow to grace a diadem. Look at that hanging lower lip! what a huge sulk and pout is there! a sulk and pout, by carpenter's measurement, about twenty feet long and five feet deep; a sulk and pout that will yield you some 500 gallons of oil and more.

A great pity, now, that this unfortunate whale should be hare-lipped. Over this lip, as over a slippery threshold, we now slide into the mouth. Good Lord! is this the road that Jonah went? The roof is about twelve feet high, and runs to a pretty sharp angle, as if there were a regular ridge-pole there; while these ribbed, arched, hairy sides, present us with those wondrous, half vertical, scimetar-shaped slats of whale-bone, say three hundred on a side, which form those Venetian blinds which have elsewhere been cursorily mentioned. The edges of these bones are fringed with hairy fibres, through which the

"A Ship on the North-West Coast Cutting in Her Last Right Whale," 1848. As it is hoisted on deck, the upper jaw of the right whale reveals a thicket of baleen. Unlike sperm whales, right whales have no teeth. Instead, the roofs of their mouths are draped with bunched rows of long keratin strips. By swimming with their mouths partially open, the whales use their baleen forests to strain krill, crustaceans and other organisms from the seawater. The two vessels depicted are actually different views of the same ship, evident by the porpoise tail affixed to the end of the jib boom. Engraved in France, this lithograph is considered Benjamin Russell's finest quality print.

New Bedford Whaling Museum

Right Whale strains the water, and in whose intricacies he retains the small fish, when open-mouthed he goes through the seas of brit in feeding time.

In old times, there seem to have prevailed the most curious fancies concerning these blinds. One voyager in Purchas calls them the wondrous 'whiskers" inside of the whale's mouth; another, 'hogs" bristles;" a third old gentleman in Hackluyt uses the following elegant language: 'There are about two hundred and fifty fins growing on each side of his upper chop, which arch over his tongue on each side of his mouth."

As every one knows, these same "hogs' bristles," "fins," "whiskers," "blinds," or whatever you please, furnish to the ladies their busks and other stiffening contrivances. But in this particular, the demand has long been on the decline. It was in Queen Anne's time that the bone was in its glory, the farthingale being then all the fashion. And as those ancient dames moved about gaily, though in the jaws of the whale, do we nowadays fly under the same jaws for protection; the umbrella being a tent spread over the same bone.

Ere this, you must have plainly seen the truth of what I started with—that the sperm whale and the right whale have almost entirely different heads. To sum up,

Drying baleen with the Arctic whaling fleet, 1887. Baleen was used to make corset stays, umbrella ribs, fishing rods, buggy whips, carriage springs and other items requiring pliability and strength.
— Photograph by Herbert L Aldrich. New Bedford Free Public Library.

then: in the right whale's there is no great well of sperm; no ivory teeth at all; no long, slender mandible of a lower jaw, like the sperm whale's. Nor in the sperm whale are there any of those blinds of bone; no huge lower lip; and scarcely anything of a tongue. The right whale has two external spout-holes, the sperm whale only one.

Look your last, now, on these venerable hooded heads, while they yet lie together; one will soon sink, unrecorded, in the sea; the other will not be very long in following.

The dying right whale, 1841. "See that amazing lower lip, pressed by accident against the vessel's side, so as firmly to embrace the jaw. Does not this whole head seem to speak of an enormous practical resolution in facing death?" — Chapter 75. Watercolor by John F. Martin aboard Lucy Ann of Wilmington, DE.

New Bedford Whaling Museum

Chapter LXXVI

The Battering-Ram

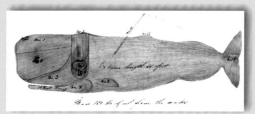

Diagram of sperm whale's yield, 1843. — Pencil and ink drawing from the journal of Joseph Bogart Hershey aboard the schooner Esquimaux of Provincetown.

Ere quitting the sperm whale's head, I would have you investigate it now with the sole view of forming to yourself some unexaggerated, intelligent estimate of whatever battering-ram power may be lodged there.

I have described how the blubber wraps the body of the whale, as the rind wraps an orange. About the head this envelope, though not so thick, is of a boneless toughness, inestimable by any man who has not handled it. The severest pointed harpoon, the sharpest lance darted by the strongest human arm, impotently rebounds from it. It is as though the forehead of the sperm whale were paved with horses' hoofs. I do not think that any sensation lurks in it.

Spermacetti whale, 1798. An early rendering of the ferocious whale. — Watercolor by Thomas Wetling aboard the ship William of London. Kendall Collection, NBWM.

Now, mark. Unerringly impelling this dead, impregnable, uninjurable wall, and this most buoyant thing within; there swims behind it all a mass of tremendous life; and all obedient to one volition, as the smallest insect. So that when I shall hereafter detail to you all the specialities and concentrations of potency everywhere lurking in this expansive monster; when I shall show you some of his more inconsiderable braining feats; I trust you will have renounced all ignorant incredulity, and be ready to abide by this; that though the Sperm Whale stove a passage through the Isthmus of Darien, and mixed the Atlantic with the Pacific, you would not elevate one hair of your eye-brow. For unless you own the whale, you are but a provincial and sentimentalist in Truth.

Stove boat, 1846-50. — Pen and watercolor from the journal of Joseph Peters aboard the ship Congaree of New Bedford.

Kendall Collection, NBWM

Chapter LXXVII

The Great Heidelburgh Tun

Sperm whale oil.
— *On exhibit at New Bedford Whaling Museum.*

Now comes the Baling of the Case. But to comprehend it aright, you must know something of the curious internal structure of the thing operated upon.

Regarding the sperm whale's head as a solid oblong, you may, on an inclined plane, sideways divide it into two quoins, whereof the lower is the bony structure, forming the cranium and jaws, and the upper an unctuous mass wholly free from bones; its broad forward end forming the expanded vertical apparent forehead of the whale. At the middle of the forehead horizontally subdivide this upper quoin, and then you have two almost equal parts, which before were naturally divided by an internal wall of a thick tendinous substance.

The lower subdivided part, called the junk, is one immense honeycomb of oil, formed by the crossing and re-crossing, into ten thousand infiltrated cells, of tough elastic white fibres throughout its whole extent. The upper part, known as the case, may be regarded as the great Heidelburgh Tun of the sperm whale. The tun of the whale contains by far the most precious of all his oily vintages; namely, the highly-prized spermaceti, in its absolutely pure, limpid, and odoriferous state. Nor is this precious substance found unalloyed in any other part of the creature. Though in life it remains perfectly fluid, yet, upon exposure to the air, after death, it soon begins to concrete; sending forth beautiful crystalline shoots, as when the first thin delicate ice is just forming in water. A large whale's case generally yields about five hundred gallons of sperm, though from unavoidable circumstances, considerable of it is spilled, leaks, and dribbles away, or is otherwise irrevocably lost in the ticklish business of securing what you can.

The Heidelburgh Tun of the sperm whale embraces the entire length of the entire top of the head; and since—as has been elsewhere set forth—the head embraces one third of the whole length of the creature, then setting that length down at eighty feet for a good sized whale, you have more than twenty-six feet for the depth of the tun, when it is lengthwise hoisted up and down against a ship's side.

As in decapitating the whale, the operator's instrument is brought close to the spot where an entrance is subsequently forced into the spermaceti magazine; he has, therefore, to be uncommonly heedful, lest a careless, untimely stroke should invade the sanctuary and wastingly let out its invaluable contents. It is this decapitated end of the head, also, which is at last elevated out of the water, and retained in that position by the enormous cutting tackles, whose hempen combinations, on one side, make quite a wilderness of ropes in that quarter.

Thus much being said, attend now, I pray you, to that marvellous and—in this particular instance—almost fatal operation whereby the sperm whale's great Heidelburgh Tun is tapped.

Hauling the case, or "Heidelburgh Tun," on the John R. Manta, 1917.

New Bedford Whaling Museum

Bailing out the case on the John R. Manta, 1925. When the oil is first exposed, it is fluid and light gold; it congeals when exposed to air.

William R. Hegarty Collection

Chapter LXXVIII

Cistern and Buckets

Cutting in a sperm whale, 1865.
— *Watercolor by Henry C. Hayward aboard bark* Awashonks *of New Bedford.*

Nimble as a cat, Tashtego mounts aloft; and without altering his erect posture, runs straight out upon the overhanging main-yard-arm, to the part where it exactly projects over the hoisted tun. He has carried with him a light tackle called a whip, consisting of only two parts, travelling through a single-sheaved block. Securing this block, so that it hangs down from the yard-arm, he swings one end of the rope, till it is caught and firmly held by a hand on deck. Then, hand-over-hand, down the other part, the Indian drops through the air, till dexterously he lands on the summit of the head. There—still high elevated above the rest of the company, to whom he vivaciously cries—he seems some Turkish Muezzin calling the good people to prayers from the top of a tower. A short-handled sharp spade being sent up to him, he diligently searches for the proper place to begin breaking into the tun. In this business he proceeds very heedfully, like a treasure-hunter in some old house, sounding the walls to find where the gold is masoned in. By the time this cautious search is over, a stout iron-bound bucket, precisely like a well-bucket, has been attached to one end of the whip; while the other end, being stretched across the deck, is there held by two or three alert hands. These last now hoist the bucket within grasp of the Indian, to whom another person has reached up a very long pole. Inserting this pole into the bucket, Tashtego downward guides the bucket into the tun, till it entirely disappears; then giving the word to the seamen at the whip, up comes the bucket again, all bubbling like a dairy-maid's pail of new milk.

Now, the people of the Pequod had been baling some time in this way; several tubs had been filled with the fragrant sperm; when all at once a queer accident happened. My God! poor Tashtego—like the twin reciprocating bucket in a veritable well, dropped head-foremost down into this great Tun of Heidelburgh, and with a horrible oily gurgling, went clean out of sight!

"Man overboard!" cried Daggoo, who amid the general consternation first came to his senses. "Swing the bucket this way!" and putting one foot into it, so as the better to secure his slippery hand-hold on the whip itself, the hoisters ran him high up to the top of the head, almost before Tashtego could have reached its interior bottom.

At this instant, while Daggoo, on the summit of the head, was clearing the whip—which had somehow got foul of the great cutting tackles—a sharp cracking noise was heard; and to the unspeakable horror of all, one of the two enormous hooks suspending the head tore out, and with a vast vibration the enormous mass sideways swung, till the drunk ship reeled and shook as if smitten

Cutting in a sperm whale, 1846-50. *This scene from Joseph Bogart Hershey's journal was drawn during his three voyages. Hershey served as first mate for one voyage of the bark* Samuel and Thomas *of Provincetown, and Captain for two voyages of the schooner* Shylock *of Provincetown.*

Kendall Collection, NBWM

by an iceberg. The one remaining hook, upon which the entire strain now depended, seemed every instant to be on the point of giving way; an event still more likely from the violent motions of the head.

"Come down, come down!" yelled the seamen to Daggoo, but with one hand holding on to the heavy tackles, so that if the head should drop, he would still remain suspended; the negro having cleared the foul line, rammed down the bucket into the now collapsed well, meaning that the buried harpooneer should grasp it, and so be hoisted out.

"Stand clear of the tackle!" cried a voice like the bursting of a rocket.

Almost in the same instant, with a thunder-boom, the enormous mass dropped into the sea; the suddenly relieved hull rolled away from it, to far down her glittering copper; and all caught their breath, as half swinging—now over the sailors' heads, and now over the water—Daggoo through a thick mist of spray, was dimly beheld clinging to the pendulous tackles, while poor, buried-alive Tashtego was sinking utterly down to the bottom of the sea! But hardly had the blinding vapor cleared away, when a naked figure with a boarding-sword in its hand, was for one swift moment seen hovering over the bulwarks. The next, a loud splash announced that my brave Queequeg had dived to the rescue. One packed rush was made to the side, and every eye counted every ripple, as moment followed moment, and no sign of either the sinker or the diver could be seen. Some hands now jumped into a boat alongside, and pushed a little off from the ship.

"Ha! ha!" cried Daggoo, all at once, from his now quiet, swinging perch overhead; and looking further off from the side, we saw an arm thrust upright from the blue waves; a sight strange to see, as an arm thrust forth from the grass over a grave.

"Both! both!—it is both!—" cried Daggoo again with a joyful shout; and soon after, Queequeg was seen boldly striking out with one hand, and with the other clutching the long hair of the Indian.

Now, how had this noble rescue been accomplished? Why, diving after the slowly descending head, Queequeg with his keen sword had made side lunges near its bottom, so as to scuttle a large hole there; then dropping his sword, had thrust his long arm far inwards and upwards, and so hauled out our poor Tash by the head.

And thus, through the courage and great skill in obstetrics of Queequeg, the deliverance, or rather, delivery of Tashtego, was successfully accomplished, in the teeth, too, of the most untoward and apparently hopeless impediments; which is a lesson by no means to be forgotten. Midwifery should be taught in the same course with fencing and boxing, riding and rowing.

Had Tashtego perished, it had been a very precious perishing; smothered in the very whitest and daintiest of fragrant spermaceti; coffined, hearsed, and tombed in the secret inner chamber and sanctum sanctorum of the whale.

"A View of Cutting In the Whale," circa 1780. — *Watercolors by Thomas Wetling aboard the ship* William *of London.*

Kendall Collection, NBWM

Chapter LXXXI

The Pequod Meets the Virgin

"Voyage Nearly Ended," circa 1867.
— *Watercolor by Benjamin Russell. Forbes Collection of the MIT Museum.*

The predestinated day arrived, and we duly met the ship Jungfrau, Derick De Deer, master, of Bremen.

For some reason, the Jungfrau seemed quite eager to pay her respects. While yet some distance from the Pequod, she rounded to, and dropping a boat, her captain was impelled towards us, impatiently standing in the bows instead of the stern.

"What has he in his hand there?" cried Starbuck, pointing to something wavingly held by the German. "Impossible!—a lamp-feeder!"

However curious it may seem for an oil-ship to be borrowing oil on the whale-ground, and however much it may invertedly contradict the old proverb about carrying coals to Newcastle, sometimes such a thing really happens.

As he mounted the deck, Ahab abruptly accosted him, without at all heeding what he had in his hand; but in his broken lingo, the German soon evinced his complete ignorance of the White Whale; immediately turning the conversation to his lamp-feeder and oil can, with some remarks touching his having to turn into his hammock at night in profound darkness—his last drop of Bremen oil being gone, and not a single flying-fish yet captured to supply the deficiency; concluding by hinting that his ship was indeed what in the Fishery is technically called a *clean* one (that is, an empty one), well deserving the name of Jungfrau or the Virgin.

His necessities supplied, Derick departed; but he had not gained his ship's side, when whales were almost simultaneously raised from the mast-heads of both vessels; and so eager for the chase was Derick, that without pausing to put his oil-can and lamp-feeder aboard, he slewed round his boat and made after the leviathan lamp-feeders.

Now, the game having risen to leeward, he and the other three German boats that soon followed him, had considerably the start of the Pequod's keels. There were eight whales, an average pod. Aware of their danger, they were going all abreast with great speed straight before the wind, rubbing their flanks as closely as so many spans of horses in harness. They left a great, wide wake, as though continually unrolling a great wide parchment upon the sea.

"Taking up the Boats," circa 1865. — *Watercolor by Benjamin Russell.*

Forbes Collection of the MIT Museum

Full in this rapid wake, and many fathoms in the rear, swam a huge, humped old bull, which by his comparatively slow progress, as well as by the unusual yellowish incrustations overgrowing him, seemed afflicted with the jaundice, or some other infirmity. His spout was short, slow, and laborious; coming forth with a choking sort of gush, and spending itself in torn shreds, followed by strange subterranean commotions in him, which seemed to have egress at his other buried extremity, causing the waters behind him to upbubble.

As an overladen Indiaman bearing down the Hindostan coast with a deck load of frightened horses, careens, buries, rolls, and wallows on her way; so did this old whale heave his aged bulk, and now and then partly turning over on his cumbrous rib-ends, expose the cause of his devious wake in the unnatural stump of his starboard fin. Whether he had lost that fin in battle, or had been born without it, it were hard to say.

With one intent all the combined rival boats were pointed for this one fish, because not only was he the largest, and therefore the most valuable whale, but he was nearest to them, and the other whales were going with such great velocity, moreover, as almost to defy pursuit for the time. At this juncture, the Pequod's keel had shot by the three German boats last lowered; but from the great start he had had, Derick's boat still led the chase, though every moment neared by his foreign rivals. The only thing they feared, was, that from being already so nigh to his mark, he would be enabled to dart his iron before they could completely overtake and pass him. As for Derick, he seemed quite confident that this would be the case, and occasionally with a deriding gesture shook his lamp-feeder at the other boats.

"The ungracious and ungrateful dog!" cried Starbuck; "he mocks and dares me with the very poor-box I filled for him not five minutes ago!"—then in his old intense whisper—"give way, greyhounds! Dog to it!"

At this moment Derick was in the act of pitching his lamp-feeder at the advancing boats, and also his oil-can; perhaps with the double view of retarding his rivals' way, and at the same time economically accelerating his own by the momentary impetus of the backward toss.

"A pod of sperm whales," circa 1862. — *Oil painting by J. R. Winn, American whaleman.*

Kendall Collection, NBWM

Sperm whaling off São Miguel, Azores, 1845. — *From Russell and Purrington's panorama. New Bedford Whaling Museum.*

Fiercely, but evenly incited by the taunts of the German, the Pequod's three boats now began ranging almost abreast; and, so disposed, momentarily neared him.

But so decided an original start had Derick had, that spite of all their gallantry, he would have proved the victor in this race, had not a righteous judgment descended upon him in a crab which caught the blade of his midship oarsman. That was a good time for Starbuck, Stubb, and Flask. An instant more, and all four boats were diagonally in the whale's immediate wake, while stretching from them, on both sides, was the foaming swell that he made.

It was a terrific, most pitiable, and maddening sight. The whale was now going head out, and sending his spout before him in a continual tormented jet; while his one poor fin beat his side in an agony of fright. Now to this hand, now to that, he yawed in his faltering flight, and still at every billow that he broke, he spasmodically sank in the sea, or sideways rolled towards the sky his one beating fin.

Seeing now that but a very few moments more would give the Pequod's boats the advantage, and rather than be thus foiled of his game, Derick chose to hazard what to him must have seemed a most unusually long dart, ere the last chance would for ever escape.

Bark President, captained by Benjamin A. Gifford, whaling in the Atlantic, 1865. — *Watercolor by Benjamin Russell,.*

Forbes Collection of the MIT Museum

But no sooner did his harpooneer stand up for the stroke, than all three tigers—Queequeg, Tashtego, Daggoo—instinctively sprang to their feet, and standing in a diagonal row, simultaneously pointed their barbs; and darted over the head of the German harpooneer, their three Nantucket irons entered the whale. Blinding vapors of foam and white-fire!

But the monster's run was a brief one. Giving a sudden gasp, he tumultuously sounded. With a grating rush, the three lines flew round the loggerheads with such a force as to gouge deep grooves in them; the harpooners, using all their dexterous might, caught repeated smoking turns with the rope to hold on; till at last—owing to the perpendicular strain from the lead-lined chocks of the boats, whence the three ropes went straight down into the blue—the gunwales of the bows were almost even with the water, while the three sterns tilted high in the air. And the whale soon ceasing to sound, for some time they remained in that attitude, fearful of expending more line, though the position was a little ticklish. But though boats have been taken down and lost in this way, yet it is this "holding on," as it is called; this hooking up by the sharp barbs of his live flesh from the back; this it is that often torments the leviathan into soon rising again to meet the sharp lance of his foes.

As the three boats lay there on that gently rolling sea, gazing down into its eternal blue noon; and as not a single groan or cry of any sort, nay, not so much as a ripple or a bubble came up from its depths; what landsman would have thought, that beneath all that silence and placidity, the utmost monster of the seas was writhing and wrenching in agony!

"Stand by, men; he stirs," cried Starbuck, as the three lines suddenly vibrated in the water, distinctly conducting upwards to them, as by magnetic wires, the life and death throbs of the whale, so that every oarsman felt them in his seat. The next moment, relieved in a great part from the downward strain at the bows, the boats gave a sudden bounce upwards, as a small ice-field will, when a dense herd of white bears are scared from it into the sea.

"Haul in! Haul in!" cried Starbuck again; "he's rising."

The lines, of which, hardly an instant before, not one hand's breadth could have been gained, were now in long quick coils flung back all dripping into the boats, and soon the whale broke water within two ship's lengths of the hunters. His motions plainly denoted his extreme exhaustion.

As the boats now more closely surrounded him, the whole upper part of his form, with much of it that is

"The Spermacetti Whale," 1837. *This scene, whcih first appeared in Hamilton's* The Natural History of the Ordinary Cetacea or Whales, *in 1837 is one of the most widely reproduced, republished and emulated images of the hunted sperm whale.* — Drawing by Stewart, engraving by W. Lizars. From Hamilton.

New Bedford Whaling Museum

ordinarily submerged, was plainly revealed. His eyes, or rather the places where his eyes had been, were beheld. As strange misgrown masses gather in the knot-holes of the noblest oaks when prostrate, so from the points which the whale's eyes had once occupied, now protruded blind bulbs, horribly pitiable to see. But pity there was none. For all his old age, and his one arm, and his blind eyes, he must die the death and be murdered, in order to light the gay bridals and other merry-makings of men, and also to illuminate the solemn churches that preach unconditional inoffensiveness by all to all. Still rolling in his blood, at last he partially disclosed a strangely discolored bunch or protuberance, the size of a bushel, low down on the flank.

"A nice spot," cried Flask; "just let me prick him there once."

"Avast!" cried Starbuck, "there's no need of that!"

But humane Starbuck was too late. At the instant of the dart an ulcerous jet shot from this cruel wound, and goaded by it into more than sufferable anguish, the whale now spouting thick blood, with swift fury blindly darted at the craft, bespattering them and their glorying crews all over with showers of gore, capsizing Flask's boat and marring the bows. It was his death stroke. For, by this time, so spent was he by loss of blood, that he helplessly rolled away from the wreck he had made; lay panting on his side, impotently flapped with his stumped fin, then over and over slowly revolved like a waning world; turned up the white secrets of his belly; lay like a log, and died. It was most piteous, that last expiring spout.

By very heedful management, when the ship drew nigh, the whale was transferred to her side, and was strongly secured there by the stiffest fluke-chains, for it was plain that unless artificially upheld, the body would at once sink to the bottom.

It so chanced that almost upon first cutting into him with the spade, the entire length of a corroded harpoon was found imbedded in his flesh, on the lower part of the bunch before described. But as the stumps of harpoons are frequently found in the dead bodies of captured whales, with the flesh perfectly healed around them, and no prominence of any kind to denote their place; therefore, there must needs have been some other unknown reason in the present case fully to account for the ulceration alluded to. But still more curious was the

Sag Harbor whalemen harpooning a sperm whale, circa 1869. — *Oil painting by William Heysman Overend.*

Kendall Collection, NBWM

fact of a lance-head of stone being found in him, not far from the buried iron, the flesh perfectly firm about it. Who had darted that stone lance? And when? It might have been darted by some Nor' West Indian long before America was discovered.

What other marvels might have been rummaged out of this monstrous cabinet there is no telling. But a sudden stop was put to further discoveries, by the ship's being unprecedentedly dragged over sideways to the sea, owing to the body's immensely increasing tendency to sink. So low had the whale now settled that the submerged ends could not be at all approached, while every moment whole tons of ponderosity seemed added to the sinking bulk, and the ship seemed on the point of going over.

"Hold on, hold on, won't ye?" cried Stubb to the body, "don't be in such a devil of a hurry to sink! Run one of ye for a prayer book and a pen-knife, and cut the big chains."

"Knife? Aye, aye," cried Queequeg, and seizing the carpenter's heavy hatchet, he leaned out of a porthole, and steel to iron, began slashing at the largest fluke-chains. But a few strokes, full of sparks, were given, when the exceeding strain effected the rest. With a terrific snap, every fastening went adrift; the ship righted, the carcase sank.

Now, this occasional inevitable sinking of the recently killed sperm whale is a very curious thing; nor has any fisherman yet adequately accounted for it. Usually the dead sperm whale floats with great buoyancy, with its side or belly considerably elevated above the surface. If the only whales that thus sank were old, meager, and broken-hearted creatures; then you might with some reason assert that this sinking is caused by an uncommon specific gravity in the fish so sinking. But it is not so, for young whales, in the highest health, and swelling with noble aspirations, prematurely cut off in the warm flush and May of life, with all their panting lard about them; even these brawny, buoyant heroes do sometimes sink.

Be it said, however, that the Sperm Whale is far less liable to this accident than any other species. But there are instances where, after the lapse of many hours or several days, the sunken whale again rises, more buoyant than in life. Gases are generated in him; he swells to a prodigious magnitude; becomes a sort of animal balloon. A line-of-battle ship could hardly keep him under then.

"The Head of a large Whale in the Agonies of Death," 1825. In this colored aquatint entitled "South Sea Whale Fishery," whaleboats from the *Amelia Wilson* and *Castor* make their kill off Buru Island, east of Suluwesi, Indonesia, a rich sperm whaling ground. In 1834, the artist, William John Huggins, was chosen by King William IV as his official marine painter. — Digitally colored.

Kendall Collection, NBWM

Chapter LXXXIV

Pitchpoling

Pitchpoling. — Ink and pencil drawings on paper by E. C. Snow, veteran of the whaling bark Abraham Baker *(1871-75).*

Of all the wondrous devices and dexterities, the sleights of hand and countless subtleties, to which the veteran whaleman is so often forced, none exceed that fine manoeuvre with the lance called pitchpoling. Small sword, or broad sword, in all its exercises boasts nothing like it. It is only indispensable with an inveterate running whale; its grand fact and feature is the wonderful distance to which the long lance is accurately darted from a violently rocking, jerking boat, under extreme headway. Steel and wood included, the entire spear is some ten or twelve feet in length; the staff is much slighter than that of the harpoon, and also of a lighter material—pine. It is furnished with a small rope called a warp, of considerable length, by which it can be hauled back to the hand after darting.

But before going further, it is important to mention here, that though the harpoon may be pitchpoled in the same way with the lance, yet it is seldom done; and when done, is still less frequently successful, on account of the greater weight and inferior length of the harpoon as compared with the lance, which in effect become serious drawbacks. As a general thing, therefore, you must first get fast to a whale, before any pitchpoling comes into play.

Look now at Stubb. Look at him; he stands upright in the tossed bow of the flying boat; wrapt in fleecy foam, the towing whale is forty feet ahead. Handling the long lance lightly, glancing twice or thrice along its length to see if it be exactly straight, Stubb whistlingly gathers up the coil of the warp in one hand, so as to secure its free end in his grasp, leaving the rest unobstructed. Then holding the lance full before his waistband's middle, he levels it at the whale; when, covering him with it, he steadily depresses the butt-end in his hand, thereby elevating the point till the weapon stands fairly balanced upon his palm, fifteen feet in the air. He minds you somewhat of a juggler, balancing a long staff on his chin. Next moment with a rapid, nameless impulse, in a superb lofty arch the bright steel spans the foaming distance, and quivers in the life spot of the whale. Instead of sparkling water, he now spouts red blood.

The agonized whale goes into his flurry; the tow-line is slackened, and the pitchpoler dropping astern, folds his hands, and mutely watches the monster die.

Whale in a flurry, circa 1850. — *Polychrome scrimshaw plaque by W. L. Roderick, surgeon on bark* Adventure *of London.*

Kendall Collection, NBWM

Chapter LXXXV

The Fountain

"Right whale staving a boat." — *Delano:* Wanderings....

That for six thousand years—and no one knows how many millions of ages before—the great whales should have been spouting all over the sea, and sprinkling and mistifying the gardens of the deep, as with so many sprinkling or mistifying pots; and that for some centuries back, thousands of hunters should have been close by the fountain of the whale, watching these sprinklings and spoutings—that all this should be, and yet, that down to this blessed minute (fifteen and a quarter minutes past one o'clock p.m. of this sixteenth day of December, A.D. 1851), it should still remain a problem, whether these spoutings are, after all, really water, or nothing but vapor—this is surely a noteworthy thing.

But owing to the mystery of the spout—whether it be water or whether it be vapor—no absolute certainty can as yet be arrived at on this head.

You have seen him spout; then declare what the spout is; can you not tell water from air? And as for this whale spout, you might almost stand in it, and yet be undecided as to what it is precisely. Nor is it at all prudent for the hunter to be over curious touching the precise nature of the whale spout. It will not do for him to be peering into it, and putting his face in it. For even when coming into slight contact with the outer, vapory shreds of the jet, which will often happen, your skin will feverishly smart, from the acridness of the thing so touching it. And I know one, who coming into still closer contact with the spout, whether with some scientific object in view, or otherwise, I cannot say, the skin peeled off from his cheek and arm. Wherefore, among whalemen, the spout is deemed poisonous; they try to evade it. Another thing; I have heard it said, and I do not much doubt it, that if the jet is fairly spouted into your eyes, it will blind you. The wisest thing the investigator can do then, it seems to me, is to let this deadly spout alone.

And how nobly it raises our conceit of the mighty, misty monster, to behold him solemnly sailing through a calm tropical sea; his vast, mild head overhung by a canopy of vapor, engendered by his incommunicable contemplations, and that vapor—as you will sometimes see it—glorified by a rainbow, as if Heaven itself had put its seal upon his thoughts.

Whales in a flurry, 1834-38. On the 12th of August at 4 PM, the crew on the ship *Canton* of New Bedford lowered four boats, got one stove and killed one whale. They returned to the ship with their whale at 9 PM. — *From the journal of David Wordel aboard ship* Canton.

Kendall Collection, NBWM

Chapter LXXXVI

The Tail

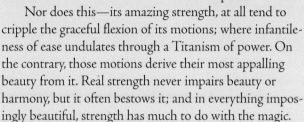

Flukes of a sperm whale, circa 1922.
— From "Down to the Sea in Ships."
New Bedford Whaling Museum.

Other poets have warbled the praises of the soft eye of the antelope, and the lovely plumage of the bird that never alights; less celestial, I celebrate a tail.

In no living thing are the lines of beauty more exquisitely defined than in the crescentic borders of these flukes. At its utmost expansion in the full grown whale, the tail will considerably exceed twenty feet across.

The entire member seems a dense webbed bed of welded sinews; but cut into it, and you find that three distinct strata compose it:—upper, middle, and lower. The fibres in the upper and lower layers, are long and horizontal; those of the middle one, very short, and running crosswise between the outside layers. This triune structure, as much as anything else, imparts power to the tail.

But as if this vast local power in the tendinous tail were not enough, the whole bulk of the leviathan is knit over with a warp and woof of muscular fibres and filaments, which passing on either side the loins and running down into the flukes, insensibly blend with them, and largely contribute to their might; so that in the tail the confluent measureless force of the whole whale seems concentrated to a point.

Nor does this—its amazing strength, at all tend to cripple the graceful flexion of its motions; where infantileness of ease undulates through a Titanism of power. On the contrary, those motions derive their most appalling beauty from it. Real strength never impairs beauty or harmony, but it often bestows it; and in everything imposingly beautiful, strength has much to do with the magic.

Such is the subtle elasticity of the organ I treat of, that whether wielded in sport, or in earnest, or in anger, whatever be the mood it be in, its flexions are invariably marked by exceeding grace. Therein no fairy's arm can transcend it.

The more I consider this mighty tail, the more do I deplore my inability to express it. At times there are gestures in it, which, though they would well grace the hand of man, remain wholly inexplicable.

"The Other End of the Whale," 1877. *A scene from Charles S. Raleigh's 197-foot "Panorama of a Whaling Voyage in the Ship* Niger."

Kendall Collection, NBWM

Chapter LXXXVII

The Grand Armada

The long and narrow peninsula of Malacca, extending south-eastward from the territories of Birmah, forms the most southerly point of all Asia. In a continuous line from that peninsula stretch the long islands of Sumatra, Java, Bally, and Timor; which, with many others, form a vast mole, or rampart, lengthwise connecting Asia with Australia, and dividing the long unbroken Indian ocean from the thickly studded oriental archipelagoes. This rampart is pierced by several sally-ports for the convenience of ships and whales; conspicuous among which are the straits of Sunda and Malacca. By the straits of Sunda, chiefly, vessels bound to China from the west, emerge into the China seas.

With a fair, fresh wind, the Pequod was now drawing nigh to these straits; Ahab purposing to pass through them into the Javan sea, and thence, cruising northwards, over waters known to be frequented here and there by the sperm whale, sweep inshore by the Philippine Islands, and gain the far coast of Japan, in time for the great whaling season there. By these means, the circumnavigating Pequod would sweep almost all the known sperm whale cruising grounds of the world, previous to descending upon the Line in the Pacific; where Ahab, though everywhere else foiled in his pursuit, firmly counted upon giving battle to Moby Dick, in the sea he was most known to frequent; and at a season when he might most reasonably be presumed to be haunting it.

Now, as many sperm whales had been captured off the western coast of Java, in the near vicinity of the straits of Sunda; indeed, as most of the ground, roundabout, was generally recognised by the fishermen as an excellent spot for cruising; therefore, as the Pequod gained more and more upon Java Head, the look-outs were repeatedly hailed, and admonished to keep wide awake. But though the green palmy cliffs of the land soon loomed on the starboard bow, and with delighted nostrils the fresh cinnamon was snuffed in the air, yet not a single jet was descried. Almost renouncing all thought of falling in with any game hereabouts, the ship had well nigh entered the straits, when the customary cheering cry was heard from aloft, and ere long a spectacle of singular magnificence saluted us.

But here be it premised, that owing to the unwearied activity with which of late they have been hunted over all four oceans, the Sperm Whales, instead of almost invari-

Whaling on the Northwest Coast. — *From Russell and Purrington's Panorama.*

ably sailing in small detached companies, as in former times, are now frequently met with in extensive herds, sometimes embracing so great a multitude, that it would almost seem as if numerous nations of them had sworn solemn league and covenant for mutual assistance and protection. To this aggregation of the Sperm Whale into such immense caravans, may be imputed the circumstance

Cruising for whales, 1841. *The ship* Lucy Ann *of Wilmington, Delaware, cruises the Pacific grounds.* — *From the journal of John F. Martin.*

Kendall Collection, NBWM

that even in the best cruising grounds, you may now sometimes sail for weeks and months together, without being greeted by a single spout; and then be suddenly saluted by what sometimes seems thousands on thousands.

Broad on both bows, at the distance of some two or three miles, and forming a great semicircle, embracing one half of the level horizon, a continuous chain of whale-jets were up-playing and sparkling in the noon-day air.

As marching armies approaching an unfriendly defile in the mountains, accelerate their march, all eagerness to place that perilous passage in their rear, and once more expand in comparative security upon the plain; even so did this vast fleet of whales now seem hurrying forward through the straits; gradually contracting the wings of their semicircle, and swimming on, in one solid, but still crescentic centre.

Crowding all sail the Pequod pressed after them; the harpooneers handling their weapons, and loudly cheering from the heads of their yet-suspended boats.

Corresponding to the crescent in our van, we beheld another in our rear. Ahab quickly revolved in his pivot-hole, crying, "Aloft there, and rig whips and buckets to wet the sails;—Malays, sir, and after us!" As if too long lurking behind the headlands, till the Pequod should fairly have entered the straits, these rascally Asiatics were now in hot pursuit, to make up for their over-cautious delay. As with glass under arm, Ahab to-and-fro paced the deck; in his forward turn beholding the monsters he chased, and in the after one the bloodthirsty pirates chasing *him*. Ahab's brow was left gaunt and ribbed, like the black sand beach after some stormy tide has been gnawing it.

When, after steadily dropping and dropping the pirates astern, the Pequod at last shot by the vivid green Cockatoo Point on the Sumatra side, emerging at last upon the broad waters beyond. But still driving on in the wake of the whales, at length they seemed abating their speed; gradually the ship neared them; and the wind now dying away, word was passed to spring to the boats.

But no sooner did the herd, by some presumed wonderful instinct of the sperm whale, become notified of the three keels that were after them,—though as yet a mile in their rear,—than they rallied again, and forming in close ranks and battalions, so that their spouts all looked like flashing lines of stacked bayonets, moved on with redoubled velocity.

Stripped to our shirts and drawers, we sprang to the white-ash, and after several hours' pulling were almost disposed to renounce the chase, when a general pausing commotion among the whales gave animating token that they were now at last under the influence of that strange perplexity of inert irresolution, which, when the fishermen perceive it in the whale, they say he is gallied. The compact martial columns in which they had been hitherto rapidly and steadily swimming, were now broken up in one measureless rout; and they seemed going mad with consternation. In all directions expanding in vast irregular circles, and aimlessly swimming hither and thither, by their short thick spoutings, they plainly betrayed their distraction of panic. This was still more strangely evinced by those of their number, who, completely paralysed as it were, helplessly floated like water-logged dismantled ships on the sea.

"Sperm Whaling with its Varieties," 1870. In this scene, Benjamin Russell depicts a variety of whaling activities. The lithograph includes captions describing scenes from to left to right: "Waiting a chance. Setting on a Whale. Ship cutting in. Just fastening. Fast Boat Changing Ends. Dead Whale Waif'd. Fast Boat Rolling up Sail. Whale sounding. Whale running. Trying out. Stove Boat. Towing Whale to Ship. Whale in a flurry (Dying)."

Spinner Collection

145

"In a School," 1874. — Davis: Nimrod of the Sea.

"Lancing," 1874. — Davis: Nimrod of the Sea.

As is customary in those cases, the boats at once separated, each making for some one lone whale on the outskirts of the shoal. In about three minutes' time, Queequeg's harpoon was flung; the stricken fish darted blinding spray in our faces, and then running away with us like light, steered straight for the heart of the herd.

Kendall Collection, NBWM

As, blind and deaf, the whale plunged forward, as if by sheer power of speed to rid himself of the iron leech that had fastened to him; as we thus tore a white gash in the sea, on all sides menaced as we flew, by the crazed creatures to and fro rushing about us; our beset boat was like a ship mobbed by ice-isles in a tempest, and striving to steer through their complicated channels and straits, knowing not at what moment it may be locked in and crushed.

All whaleboats carry certain curious contrivances, originally invented by the Nantucket Indians, called druggs. Two thick squares of wood of equal size are stoutly clenched together, so that they cross each other's grain at right angles; a line of considerable length is then attached to the middle of this block, and the other end of the line being looped, it can in a moment be fastened to a harpoon. It is chiefly among gallied whales that this drugg is used. Our boat was furnished with three of them. The first and second were successfully darted, and we saw the whales staggeringly running off, fettered by the enormous sidelong resistance of the towing drugg.

It had been next to impossible to dart these drugged-harpoons, were it not that as we advanced into the herd, our whale's way greatly diminished; moreover, that as we went still further and further from the circumference of commotion, the direful disorders seemed waning. So that when at last the jerking harpoon drew out, and the towing whale sideways vanished; then, with the tapering force of his parting momentum, we glided between two whales into the innermost heart of the shoal, as if from some mountain torrent we had slid into a serene

Inspired, 1850. *A week of adventure, danger, hard work and good fortune is recorded on this page from the journal of Warren W. Baker aboard the brig* Leonidas *of Westport, Massachusetts. On Tuesday, September 25, Baker wrote: "…Saw sperm whales toward the boats and struck a right whale…" On Thursday, October 1: "…All hands employed in stowing in…finished stowing 80 lbs of sperm oil."*

Sperm whaling scene, circa 1840. Corset busk of skeletal whale bone. — *Anonymous whaler-scrimshander. Kendall Collection, NBWM.*

valley lake. In this central expanse the sea presented that smooth satin-like surface, called a sleek, produced by the subtle moisture thrown off by the whale in his more quiet moods. Yes, we were now in that enchanted calm which they say lurks at the heart of every commotion.

But far beneath this wondrous world upon the surface, another and still stranger world met our eyes as we gazed over the side. For, suspended in those watery vaults, floated the forms of the nursing mothers of the whales, and those that by their enormous girth seemed shortly to become mothers. The lake, as I have hinted, was to a considerable depth exceedingly transparent; and as human infants while suckling will calmly and fixedly gaze away from the breast, as if leading two different lives at the time; and while yet drawing mortal nourishment, be still spiritually feasting upon some unearthly reminiscence;—even so did the young of these whales seem looking up towards us, but not at us, as if we were but a bit of Gulf-weed in their new-born sight. Floating on their sides, the mothers also seemed quietly eyeing us.

One of these little infants, that from certain queer tokens seemed hardly a day old, might have measured some fourteen feet in length, and some six feet in girth. He was a little frisky; though as yet his body seemed scarce yet recovered from that irksome position it had so lately occupied in the maternal reticule; where, tail to head, and all ready for the final spring, the unborn whale lies bent like a Tartar's bow. The delicate side-fins, and the palms of his flukes, still freshly retained the plaited crumpled appearance of a baby's ears newly arrived from foreign parts.

Meanwhile, as we thus lay entranced, the occasional sudden frantic spectacles in the distance evinced the activity of the other boats, still engaged in drugging the whales on the frontier of the host; or possibly carrying on the war within the first circle, where abundance of room and some convenient retreats were afforded them. But the sight of the enraged drugged whales now and then blindly darting to and fro across the circles, was nothing to what at last met our eyes. It is sometimes the custom when fast to a whale more than commonly powerful and alert, to seek to

The North Cape, New Zealand and Sperm Whale Fishery, 1838. — *Illustration from J. S. Polack, engraved by W. Read.*

Kendall Collection, NBWM

hamstring him, as it were, by sundering or maiming his gigantic tail-tendon. It is done by darting a short-handled cutting-spade, to which is attached a rope for hauling it back again. A whale wounded (as we afterwards learned) in this part, but not effectually, as it seemed, had broken away from the boat, carrying along with him half of the harpoon line; and in the extraordinary agony of the wound, he was now dashing among the revolving circles like the lone mounted desperado Arnold, at the battle of Saratoga, carrying dismay wherever he went.

This terrific object seemed to recall the whole herd from their stationary fright. A low advancing hum was soon heard; and then like to the tumultuous masses of block-ice when the great river Hudson breaks up in Spring, the entire host of whales came tumbling upon their inner centre, as if to pile themselves up in one common mountain. Instantly Starbuck and Queequeg changed places; Starbuck taking the stern.

The boat was now all but jammed between two vast black bulks, leaving a narrow Dardanelles between their long lengths. But by desperate endeavor we at last shot into a temporary opening; then giving way rapidly, and at the same time earnestly watching for another outlet.

After many similar hair-breadth escapes, we at last swiftly glided into what had just been one of the outer circles, but now crossed by random whales, all violently making for one centre.

Riotous and disordered as the universal commotion now was, it soon resolved itself into what seemed a systematic movement; for having clumped together at last in one dense body, they then renewed their onward flight with augmented fleetness. Further pursuit was useless; but the boats still lingered in their wake to pick up what drugged whales might be dropped astern, and likewise to secure one which Flask had killed and waifed. The waif is a pennoned pole, two or three of which are carried by every boat; and which, when additional game is at hand, are inserted upright into the floating body of a dead whale, both to mark its place on the sea, and also as token of prior possession, should the boats of any other ship draw near.

The result of this lowering was somewhat illustrative of that sagacious saying in the Fishery,—the more whales the less fish. Of all the drugged whales only one was captured. The rest contrived to escape for the time, but only to be taken, as will hereafter be seen, by some other craft than the Pequod.

"A Shoal of Sperm Whales off the Island of Hawaii, 1833." The inscription continues: "In which the ships Enterprise, Wm. Rotch, Pocahontas, & Houqua were engaged 16th Decr. 1833. To the Merchants, Capts., Officers & Crews engaged in the Whale Fishing this is respectfully inscribed by Corns. B. Hulsart who lost an Arm on board the Whale Ship Superior of New London and was on board of the Enterprise at the time." This well-known engraving, popular in its day, is believed by some Melville historians to have inspired the passage about the crippled beggar on Tower Hill in Chapter 57. It is also one of few early whaling prints depicting the Hawaiian Islands. — *Aquatint engraved by John Hill, based on a sketch by Cornelius B. Hulsart. Information: Ingalls.*

Spinner Collection

Chapter LXXXVIII

Schools and Schoolmasters

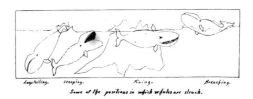

"Some of the positions in which whales are struck," 1904.
— *Ink drawing by John Bertoncini, whaling master and shipboard artist. Kendall Collection, NBWM.*

Now, though such great bodies are at times encountered, yet, as must have been seen, even at the present day, small detached bands are occasionally observed, embracing from twenty to fifty individuals each. Such bands are known as schools. They generally are of two sorts; those composed almost entirely of females, and those mustering none but young vigorous males, or bulls, as they are familiarly designated.

In cavalier attendance upon the school of females, you invariably see a male of full grown magnitude, but not old; who, upon any alarm, evinces his gallantry by falling in the rear and covering the flight of his ladies. In truth, this gentleman is a luxurious Ottoman, swimming about over the watery world, surroundingly accompanied by all the solaces and endearments of the harem.

It is very curious to watch this harem and its lord in their indolent ramblings. Like fashionables, they are for ever on the move in leisurely search of variety.

As ashore, the ladies often cause the most terrible duels among their rival admirers; just so with the whales, who sometimes come to deadly battle, and all for love. They fence with their long lower jaws, sometimes locking them together, and so striving for the supremacy like elks that warringly interweave their antlers. Not a few are captured having the deep scars of these encounters,—furrowed heads, broken teeth, scolloped fins; and in some instances, wrenched and dislocated mouths.

The schools composing none but young and vigorous males, previously mentioned, offer a strong contrast to the harem schools. For while those female whales are characteristically timid, the young males, or forty-barrel-bulls, as they call them, are by far the most pugnacious of all leviathans.

The forty-barrel-bull schools are larger than the harem schools. Like a mob of young collegians, they are full of fight, fun, and wickedness, tumbling round the world at such a reckless, rollicking rate, that no prudent underwriter would insure them any more than he would a riotous lad at Yale or Harvard. They soon relinquish this turbulence though, and when about three fourths grown, break up, and separately go about in quest of settlements, that is, harems.

Another point of difference between the male and female schools is still more characteristic of the sexes. Say you strike a Forty-barrel-bull—poor devil! all his comrades quit him. But strike a member of the harem school, and her companions swim around her with every token of concern, sometimes lingering so near her and so long, as themselves to fall a prey.

"Whaling on Japan Grounds," 1843. — *From the journal of George A. Gould aboard ship Columbia of Nantucket. Kendall Collection, NBWM.*

Chasing a school of sperm whales off Hivaoa Island, Marquesas, 1845. — *From Russell and Purrington's panorama.*

New Bedford Whaling Museum

Chapter LXXXIX

Fast-fish and Loose-fish

Fast to a sperm whale. Corset busk of skeletal whalebone, 1830s–50s.
— *By unknown scrimshander. Kendall Collection, NBWM.*

The allusion to the waifs and waif-poles necessitates some account of the laws and regulations of the whale fishery, of which the waif may be deemed the grand symbol and badge.

It frequently happens that when several ships are cruising in company, a whale may be struck by one vessel, then escape, and be finally killed and captured by another vessel; and herein are indirectly comprised many minor contingencies, all partaking of this one grand feature. For example,—after a weary and perilous chase and capture of a whale, the body may get loose from the ship by reason of a violent storm; and drifting far away to leeward, be retaken by a second whaler, who, in a calm, snugly tows it alongside, without risk of life or line. Thus the most vexatious and violent disputes would often arise between the fishermen, were there not some written or unwritten, universal, undisputed law applicable to all cases.

I. A Fast-Fish belongs to the party fast to it.

II. A Loose-Fish is fair game for anybody who can soonest catch it.

But what plays the mischief with this masterly code is the admirable brevity of it, which necessitates a vast volume of commentaries to expound it.

First: What is a Fast-Fish? Alive or dead a fish is technically fast, when it is connected with an occupied ship or boat, by any medium at all controllable by the occupant or occupants,—a mast, an oar, a nine-inch cable, a telegraph wire, or a strand of cobweb, it is all the same. Likewise a fish is technically fast when it bears a waif, or any other recognised symbol of possession; so long as the party waifing it plainly evince their ability at any time to take it alongside, as well as their intention so to do.

But if the doctrine of fast-fish be pretty generally applicable, the kindred doctrine of the loose-fish is still more widely so.

What was America in 1492 but a loose-fish, in which Columbus struck the Spanish standard by way of waifing it for his royal master and mistress? What was Poland to the Czar? What India to England? What at last will Mexico be to the United States? All Loose-Fish.

What are the Rights of Man and the Liberties of the World but Loose-Fish? What all men's minds and opinions but Loose-Fish? What is the principle of religious belief in them but a Loose-Fish? What to the ostentatious smuggling verbalists are the thoughts of thinkers but Loose-Fish? What is the great globe itself but a Loose-Fish? And what are you, reader, but a Loose-Fish and a Fast-Fish, too?

"Whaling on Japan Grounds," August 1843. As boats from different ships attack the school, their waifed whales are marked with flags. The black whaleboat (left of center) belongs to the ship Charles and Henry—*the Nantucket whaler Melville signed onto as a harpooner one month after his mutinous desertion of the* Lucy Ann. *Melville's voyage on* Charles and Henry *lasted just six months. He was discharged from the ship in May 1843, about four months before this journal entry. The* Charles and Henry *was an unlucky ship. In November 1842, Captain John Coleman wrote to the owners: "We saw whales fourteen times…and only got seven which made 140 bls., whales have been very wild…and almost impossible to get nigh them though I can assure you that I have not got the best whaleman in the world, my boat steerers have missed two hundred and fifty bls." In Melville's six months aboard* Charles and Henry, *they captured not one sperm whale. When the Captain decided to head to the Northwest Coast, Melville asked for his discharge. The ship returned to Nantucket in March 1845 with only 689 barrels of sperm oil.* Charles and Henry *was lost on her next voyage in 1845.* — *Watercolor from the journal of George A. Gould aboard the ship* Columbia *of Nantucket.*

Kendall Collection, NBWM

Chapter XCI, XCII

The Pequod Meets the Rose-bud · Ambergris

It was a week or two after the last whaling scene recounted, and when we were slowly sailing over a sleepy, vapory, mid-day sea, that a peculiar and not very pleasant smell was smelt in the sea.

Presently, the vapors in advance slid aside; and there in the distance lay a ship, whose furled sails betokened that some sort of whale must be alongside. As we glided nearer, the stranger showed French colors from his peak; and it was plain that the whale alongside must be what the fishermen call a blasted whale, that is, a whale that has died unmolested on the sea, and so floated an unappropriated corpse.

Coming still nearer with the expiring breeze, we saw that the Frenchman had a second whale alongside; and this second whale seemed even more of a nosegay than the first. In truth, it turned out to be one of those problematical whales that seem to dry up and die with a sort of prodigious dyspepsia, or indigestion; leaving their defunct bodies almost entirely bankrupt of anything like oil. Nevertheless, no knowing fisherman will ever turn up his nose at such a whale as this, however much he may shun blasted whales in general.

The Pequod had now swept so nigh to the stranger, that Stubb vowed he recognized his cutting spade-pole entangled in the lines that were knotted round the tail of one of these whales.

"There's a pretty fellow, now," he banteringly laughed, standing in the ship's bows, "there's a jackal for ye! I well know that these Crappoes of Frenchmen are but poor devils in the fishery; sometimes sailing from their port with their hold full of boxes of tallow candles, and cases of snuffers, foreseeing that all the oil they will get won't be enough to dip the Captain's wick into; but look ye, here's a Crappo that is content with our leavings, the drugged whale there, I mean; aye, and is content too with scraping the dry bones of that other precious fish he has there. Poor devil! For what oil he'll get from that drugged whale there, wouldn't be fit to burn in a jail; no, not in a condemned cell. And as for the other whale, why, I'll agree to get more oil by chopping up and trying out these three masts of ours, than he'll get from that bundle of bones; though, now that I think of it, it may contain something worth a good deal more than oil; yes, ambergris. I wonder now if our old man has thought of that. It's worth trying. Yes, I'm in for it;" and so saying he started for the quarter-deck.

Issuing from the cabin, Stubb now called his boat's crew, and pulled off for the stranger. Drawing across her bow, he perceived that in accordance with the fanciful French taste, the upper part of her stem-piece was carved in the likeness of a huge drooping stalk, was painted green, and for thorns had copper spikes projecting from it here and there; the whole terminating in a symmetrical folded bulb of a bright red color. Upon her head boards, in large gilt letters, he read "Bouton de Rose,"—Rose-button, or Rose-bud; and this was the romantic name of this aromatic ship.

"A wooden rose-bud, eh?" he cried with his hand to his nose, "that will do very well; but how like all creation it smells!"

Now in order to hold direct communication with the people on deck, he had to pull round the bows to the starboard side, and thus come close to the blasted whale; and so talk over it.

Arrived then at this spot, with one hand still to his nose, he bawled—"Bouton-de-Rose, ahoy! Are there any of you Bouton-de-Roses that speak English?"

"Yes," rejoined a Guernsey-man from the bulwarks, who turned out to be the chief-mate.

"Well, then, my Bouton-de-Rose-bud, have you seen the white whale?"

"Never heard of such a whale. Cachalot Blanche! White whale—no."

"Very good, then; good bye now, and I'll call again in a minute."

Pure sperm, circa 1870. *William F. Nye began manufacturing whale oil in 1866 and by 1877 was refining 150,000 gallons annually, becoming the world's largest manufacturer of sewing machine, watch and clock oils. Nye made a deal with publisher Charles Taber to use Taber's discountinued lithographic print "The Conflict" (p. 107) for his ad poster.*

New Bedford Free Public Library

Letterhead, circa 1859, with the popular engraving of the dying whale that originated with Hamilton and has been used countless times over. — *Kendall Collection, NBWM.*

Then rapidly pulling back towards the Pequod, and seeing Ahab leaning over the quarter-deck rail awaiting his report, he moulded his two hands into a trumpet and shouted—"No, Sir! No!" Upon which Ahab retired, and Stubb returned to the Frenchman.

He now perceived that the Guernsey-man, who had just got into the chains, and was using a cutting-spade, had slung his nose in a sort of bag.

"What's the matter with your nose, there?" said Stubb. "Broke it?"

"I wish it was broken, or that I didn't have any nose at all!" answered the Guernsey-man, who did not seem to relish the job he was at very much. "But what are you holding *yours* for?"

"Oh, nothing! It's a wax nose; I have to hold it on. Fine day, aint it? Air rather gardenny, I should say; throw us a bunch of posies, will ye, Bouton-de-Rose?"

"What in the devil's name do you want here?" roared the Guernsey-man, flying into a sudden passion.

"Oh! keep cool—cool? yes, that's the word; why don't you pack those whales in ice while you're working at 'em? But joking aside, though; do you know, Rose-bud, that it's all nonsense trying to get any oil out of such whales? As for that dried up one, there, he hasn't a gill in his whole carcase."

"I know that well enough; but, d'ye see, the Captain here won't believe it; this is his first voyage. But come aboard, and mayhap he'll believe you, if he won't me; and so I'll get out of this dirty scrape."

"Anything to oblige ye, my sweet and pleasant fellow," rejoined Stubb, and with that he soon mounted to the deck. The sailors, in tasselled caps of red worsted, were getting the heavy tackles in readiness for the whales. All their noses upwardly projected from their faces like so many jib-booms. Now and then pairs of them would drop their work, and run up to the mast-head to get some fresh air. Some thinking they would catch the plague, dipped oakum in coal-tar, and at intervals held it to their nostrils. Others having broken the stems of their pipes almost short off at the bowl, were vigorously puffing tobacco-smoke, so that it constantly filled their olfactories.

Stubb was struck by a shower of outcries and anathemas proceeding from the Captain's round-house abaft; and looking in that direction saw a fiery face thrust from behind the door, which was held ajar from within.

Marking all this, Stubb argued well for his scheme, and turning to the Guernsey-man had a little chat with him, during which the stranger mate expressed his detestation of his Captain as a conceited ignoramus, who had brought them all into so unsavory and unprofitable a pickle. The two quickly concocted a little plan for both circumventing and satirizing the Captain, without his at all dreaming of distrusting their sincerity.

By this time their destined victim appeared from his cabin. He was a small and dark, but rather delicate looking man for a sea-captain, with large whiskers and moustache, however; and wore a red cotton velvet vest with watch-seals at his side. To this gentleman, Stubb was now politely introduced by the Guernsey-man, who at once ostentatiously put on the aspect of interpreting between them.

"What shall I say to him first?" said he.

"Why," said Stubb, eyeing the velvet vest and the watch and seals, "you may as well begin by telling him that he looks a sort of babyish to me, though I don't pretend to be a judge."

"He says, Monsieur," said the Guernsey-man, in French, turning to his captain, "that only yesterday his ship spoke a vessel, whose captain and chief-mate, with six sailors, had all died of a fever caught from a blasted whale they had brought alongside."

Upon this the captain started, and eagerly desired to know more.

"What now?" said the Guernsey-man to Stubb.

"Why, since he takes it so easy, tell him that now I have eyed him carefully, I'm quite certain that he's no more fit to command a whale-ship than a St. Jago monkey. In fact, tell him from me he's a baboon."

"He vows and declares, Monsieur, that the other whale, the dried one, is far more deadly than the blasted one; in fine, Monsieur, he conjures us, as we value our lives, to cut loose from these fish."

Instantly the captain ran forward, and in a loud voice commanded his crew to desist from hoisting the cutting-tackles, and at once cast loose the cables and chains confining the whales to the ship.

"What now?" said the Guernsey-man, when the captain had returned to them.

"Why, let me see; yes, you may as well tell him I've diddled him, and (aside to himself) perhaps somebody else."

Bottle label, circa 1900. — *Kendall Collection, NBWM.*

"He says, Monsieur, that he's very happy to have been of any service to us."

Hearing this, the captain vowed that they were the grateful parties (meaning himself and mate) and concluded by inviting Stubb down into his cabin to drink a bottle of Bordeaux.

"Thank him heartily; but tell him it's against my principles to drink with the man I've diddled. In fact, tell him I must go."

"He says, Monsieur, that his principles won't admit of his drinking; but that if Monsieur wants to live another day to drink, then Monsieur had best drop all four boats, and pull the ship away from these whales, for it's so calm they won't drift."

While the Frenchman's boats, then, were engaged in towing the ship one way, Stubb benevolently towed away at his whale the other way, ostentatiously slacking out a most unusually long tow-line.

Presently a breeze sprang up; Stubb feigned to cast off from the whale; hoisting his boats, the Frenchman soon increased his distance, while the Pequod slid in between him and Stubb's whale. Whereupon Stubb quickly pulled to the floating body, and hailing the Pequod to give notice of his intentions, at once proceeded to reap the fruit of his unrighteous cunning. Seizing his sharp boat-spade, he commenced an excavation in the body, a little behind the side fin. His boat's crew were all in high excitement, eagerly helping their chief, and looking as anxious as gold-hunters.

"I have it, I have it," cried Stubb, with delight, striking something in the subterranean regions, "a purse! a purse!"

Dropping his spade, he thrust both hands in, and drew out handfuls of something that looked like ripe Windsor soap, or rich mottled old cheese; very unctuous and savory withal. You might easily dent it with your thumb; it is of a hue between yellow and ash color. And this, good friends, is ambergris, worth a gold guinea an ounce to any druggist. Some six handfuls were obtained; but more was unavoidably lost in the sea, and still more, perhaps, might have been secured were it not for impatient Ahab's loud command to Stubb to desist, and come on board, else the ship would bid them good bye.

CHAPTER CXII

Now this ambergris is a very curious substance, and so important as an article of commerce, that in 1791 a certain Nantucket-born Captain Coffin was examined at the bar of the English House of Commons on that subject.

Ambergris is soft, waxy, and so highly fragrant and spicy, that it is largely used in perfumery, pastiles, precious candles, hair-powders, and pomatum. The Turks use it in cooking, and also carry it to Mecca, for the same purpose that frankincense is carried to St. Peter's in Rome. Some wine merchants drop a few grains into claret, to flavor it.

Who would think, then, that such fine ladies and gentlemen should regale themselves with an essence found in the inglorious bowels of a sick whale! Yet so it is. By some, ambergris is supposed to be the cause, and by others the effect, of the dyspepsia in the whale.

The virtue of the whale, 1840. *A colored lithograph entitled "Graphic Illustrations of Animals. Showing Their Utility to Man..." extols the usefulness of the whale for providing man with light, food, manure, whalebone, oil, and (for commerce) ambergris and spermacetti. In this vignette, a woman is buying perfume—an ambergris-based product.*
— *Engraving by W. Hawkins.*

Kendall Collection, NBWM

Chapter XCIII

The Castaway

It was but some few days after encountering the Frenchman, that a most significant event befell the most insignificant of the Pequod's crew.

Now, in the whale ship, it is not every one that goes in the boats. Some few hands are reserved called ship-keepers, whose province it is to work the vessel while the boats are pursuing the whale. It was so in the Pequod with the little negro Pippin by nick-name, Pip by abbreviation.

It came to pass, that Stubb's after-oarsman chanced so to sprain his hand, as for a time to become quite maimed; and, temporarily, Pip was put into his place. The first time Stubb lowered with him, Pip evinced much nervousness; but happily, for that time, escaped close contact with the whale; and therefore came off not altogether discreditably.

Now upon the second lowering, the boat paddled upon the whale; and as the fish received the darted iron, it gave its customary rap, which happened, in this instance, to be right under poor Pip's seat. The involuntary consternation of the moment caused him to leap, paddle in hand, out of the boat; and in such a way, that part of the slack whale line coming against his chest, he breasted it overboard with him, so as to become entangled in it, when at last plumping into the water. That instant the stricken whale started on a fierce run, the line swiftly straightened; and presto! poor Pip came all foaming up to the chocks of the boat, remorselessly dragged there by the line, which had taken several turns around his chest and neck.

Tashtego stood in the bows. He was full of the fire of the hunt. He hated Pip for a poltroon. Snatching the boat-knife from its sheath, he suspended its sharp edge over the line, and turning towards Stubb, exclaimed interrogatively, "Cut?" Meantime Pip's blue, choked face plainly looked, Do, for God's sake! All passed in a flash. In less than half a minute, this entire thing happened.

"Damn him, cut!" roared Stubb; and so the whale was lost and Pip was saved.

So soon as he recovered himself, the poor little negro was assailed by yells and execrations from the crew. Tranquilly permitting these irregular cursings to evaporate, Stubb then in a plain, business-like, but still half humorous manner, cursed Pip officially; and that done, unofficially gave him much wholesome advice. Now, in general, *Stick to the boat*, is your true motto in whaling; but cases will sometimes happen when *Leap from the boat*, is still better. Stubb suddenly dropped all advice, and concluded with a peremptory command, "Stick to the boat, Pip, or by the Lord, I won't pick you up if you jump; mind that. We can't afford to lose whales by the likes of you; a whale would sell for thirty times what you would, Pip, in Alabama. Bear that in mind, and don't jump any more."

"Sunbeam Lowering Boats," 1906. — Painting by Clifford W. Ashley. New Bedford Free Public Library.

But we are all in the hands of the Gods; and Pip jumped again. It was under very similar circumstances to the first performance; but this time he did not breast out the line; and hence, when the whale started to run, Pip was left behind on the sea.

But had Stubb really abandoned the poor little negro to his fate? No; he did not mean to, at least. Because there were two boats in his wake, and he supposed, no doubt, that they would of course come up to Pip very quickly, and pick him up. But it so happened, that those boats, without seeing Pip, suddenly spying whales close to them on one side, turned, and gave chase; and Stubb's boat was now so far away, and he and all his crew so intent upon his fish, that Pip's ringed horizon began to expand around him miserably. By the merest chance the ship itself at last rescued him; but from that hour the little negro went about the deck an idiot; such, at least, they said he was. The sea had jeeringly kept his finite body up, but drowned the infinite of his soul.

For the rest, blame not Stubb too hardly. The thing is common in that fishery; and in the sequel of the narrative, it will then be seen what like abandonment befell myself.

"Rescue at sea," 1846. *Ten shipwrecked crew from the American brig* Somers *are plucked from the sea by an imperiled boat from the French brig* Le Mercure. — *Colored lithograph by A. Mayer.*

Kendall Collection, NBWM

Chapter XCIV

A Squeeze of the Hand

"Bailing the Case," 1905.
— *Painting by Clifford W. Ashley. New Bedford Free Public Library.*

That whale of Stubb's so dearly purchased, was duly brought to the Pequod's side, where all those cutting and hoisting operations previously detailed, were regularly gone through, even to the baling of the Heidelburgh Tun, or Case.

While some were occupied with this latter duty, others were employed in dragging away the larger tubs, so soon as filled with the sperm; and when the proper time arrived, this same sperm was carefully manipulated ere going to the try-works.

It had cooled and crystallized to such a degree, that when, with several others, I sat down before a large Constantine's bath of it, I found it strangely concreted into lumps, here and there rolling about in the liquid part. It was our business to squeeze these lumps back into fluid. A sweet and unctuous duty! No wonder that in old times this sperm was such a favorite cosmetic. Such a clearer! such a sweetener! such a softener! such a delicious mollifier! After having my hands in it for only a few minutes, my fingers felt like eels, and began, as it were, to serpentine and spiralize.

Would that I could keep squeezing that sperm for ever! For now, since by many prolonged, repeated experiences, I have perceived that in all cases man must eventually lower, or at least shift, his conceit of attainable felicity; not placing it anywhere in the intellect or the fancy; but in the wife, the heart, the bed, the table, the saddle, the fire-side, the country; now that I have perceived all this, I am ready to squeeze case eternally. In thoughts of the visions of the night, I saw long rows of angels in paradise, each with his hands in a jar of spermaceti.

Now, while discoursing of sperm, it behooves to speak of other things akin to it, in the business of preparing the sperm whale for the try-works.

First comes white-horse, so called, which is obtained from the tapering part of the fish, and also from the thicker portions of his flukes. It is tough with congealed tendons—a wad of muscle—but still contains some oil. After being severed from the whale, the white-horse is first cut into portable oblongs ere going to the mincer. They look much like blocks of Berkshire marble.

Blankets on deck, 1925. *Several tons of blubber stripped from three small whales are piled on deck of the* John R. Manta. *The strips, called blanket pieces, are laid flat with the skin side down so they will be easier to cut up.*

William R. Hegarty Collection

Plum-pudding is the term bestowed upon certain fragmentary parts of the whale's flesh, here and there adhering to the blanket of blubber. As its name imports, it is of an exceedingly rich, mottled tint, with a bestreaked snowy and golden ground, dotted with spots of the deepest crimson and purple. It is plums of rubies, in pictures of citron.

There is another substance, and a very singular one, which turns up in the course of this business, but which I feel it to be very puzzling adequately to describe. It is called slobgollion; an appellation original with the whalemen, and even so is the nature of the substance. It is an ineffably oozy, stringy affair, most frequently found in the tubs of sperm, after a prolonged squeezing, and subsequent decanting. I hold it to be the wondrously thin, ruptured membranes of the case, coalescing.

Gurry, so called, is a term properly belonging to right whalemen, but sometimes incidentally used by the sperm fishermen. It designates the dark, glutinous substance which is scraped off the back of the Greenland or right whale, and much of which covers the decks of those inferior souls who hunt that ignoble Leviathan.

Nippers. Strictly this word is not indigenous to the whale's vocabulary. But as applied by whalemen, it becomes so. A whaleman's nipper is a short firm strip of tendinous stuff cut from the tapering part of Leviathan's tail: it averages an inch in thickness, and for the rest, is about the size of the iron part of a hoe. Edgewise moved along the oily deck, it operates like a leathern squilgee.

But to learn all about these recondite matters, your best way is at once to descend into the blubber-room, and have a long talk with its inmates. This place has previously been mentioned as the receptacle for the blanket-pieces, when stript and hoisted from the whale. When the proper time arrives for cutting up its contents, this apartment is a scene of terror to all tyros, especially by night. On one side, lit by a dull lantern, a space has been left clear for the workmen. They generally go in pairs,—a pike-and-gaff-man and a spade-man. The whaling-pike is similar to a frigate's boarding-weapon of the same name. The gaff is something like a boat-hook. With his gaff, the gaffman hooks on to a sheet of blubber, and strives to hold it from slipping, as the ship pitches and lurches about. Meanwhile, the spade-man stands on the sheet itself, perpendicularly chopping it into the portable horse-pieces. This spade is sharp as hone can make it; the spademan's feet are shoeless; the thing he stands on will sometimes irresistibly slide away from him, like a sledge. If he cuts off one of his own toes, or one of his assistants', would you be very much astonished? Toes are scarce among veteran blubber-room men.

Bailing an 85-barrel case on ship **Sunbeam,** *1904.* — Kendall Collection, NBWM.

Whaling implements, 1841. Skillfully drawn parts' list for the whale-ship Lucy Ann *of Wilmington, Delaware.* — From the journal of John F. Martin.

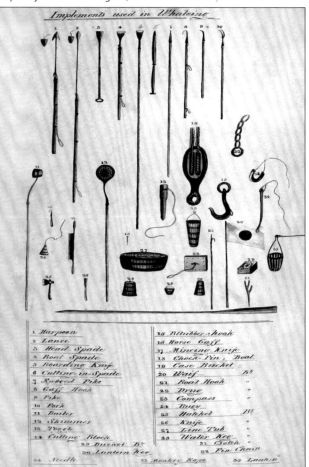

Kendall Collection, NBWM

Chapter XCV

The Cassock

"Mincers." — *Browne*: Etchings from a Whaling Cruise.

Had you stepped on board the Pequod at a certain juncture of this post-mortemizing of the whale; and had you strolled forward nigh the windlass, pretty sure am I that you would have scanned with no small curiosity a very strange, enigmatical object, which you would have seen there, lying along lengthwise in the lee scuppers. Not the wondrous cistern in the whale's huge head; not the prodigy of his unhinged lower jaw; not the miracle of his symmetrical tail; none of these would so surprise you, as half a glimpse of that unaccountable cone,—longer than a Kentuckian is tall, nigh a foot in diameter at the base, and jet-black as Yojo, the ebony idol of Queequeg. And an idol, indeed, it is; or, rather, in old times, its likeness was.

Look at the sailor, called the mincer, who now comes along, and assisted by two allies, heavily backs the grandissimus, as the mariners call it, and with bowed shoulders, staggers off with it as if he were a grenadier carrying a dead comrade from the field. Extending it upon the forecastle deck, he now proceeds cylindrically to remove its dark pelt, as an African hunter the pelt of a boa. This done he turns the pelt inside out, like a pantaloon leg; gives it a good stretching, so as almost to double its diameter; and at last hangs it, well spread, in the rigging, to dry. Ere long, it is taken down; when removing some three feet of it, towards the pointed extremity, and then cutting two slits for arm-holes at the other end, he lengthwise slips himself bodily into it. The mincer now stands before you invested in the full canonicals of his calling. Immemorial to all his order, this investiture alone will adequately protect him, while employed in the peculiar functions of his office.

That office consists in mincing the horse-pieces of blubber for the pots; an operation which is conducted at a curious wooden horse, planted endwise against the bulwarks, and with a capacious tub beneath it, into which the minced pieces drop, fast as the sheets from a rapt orator's desk. Arrayed in decent black; occupying a conspicuous pulpit; intent on bible leaves; what a candidate for an archbishoprick, what a lad for a Pope were this mincer![1]

[1] Bible leaves! Bible leaves! This is the invariable cry from the mates to the mincer. It enjoins him to be careful, and cut his work into as thin slices as possible, inasmuch as by so doing the business of boiling out the oil is much accelerated, and its quantity considerably increased, besides perhaps improving it in quality.

Making books on John R. Manta, 1925. The horse pieces are sliced like bread with a mincing knife so the strip of blubber resembles an accordian. The strip is called a book; each cut, or page, is called a bible leaf.

William R. Hegarty Collection

Chapter XCVI

The Try-works

Skimmer and ladle made of tinned steel attached to 90-inch ash poles. — *New Bedford, 19th Century. Kendall Collection, NBWM.*

Besides her hoisted boats, an American whaler is outwardly distinguished by her try-works. The try-works are planted between the foremast and main-mast, the most roomy part of the deck. The timbers beneath are of a peculiar strength, fitted to sustain the weight of an almost solid mass of brick and mortar, some ten feet by eight square, and five in height. The foundation does not penetrate the deck, but the masonry is firmly secured to the surface by ponderous knees of iron bracing it on all sides, and screwing it down to the timbers. On the flanks it is cased with wood, and at top completely covered by a large, sloping, battened hatchway. Removing this hatch we expose the great try-pots, two in number, and each of several barrels' capacity. When not in use, they are kept remarkably clean. Sometimes they are polished with soapstone and sand, till they shine within like silver punch-bowls. During the night-watches some cynical old sailors will crawl into them and coil themselves away there for a nap. While employed in polishing them—one man in each pot, side by side—many confidential communications are carried on, over the iron lips. It is a place also for profound mathematical meditation. It was in the left hand try-pot of the Pequod, with the soapstone diligently circling round me, that I was first indirectly struck by the remarkable fact, that in geometry all bodies gliding along the cycloid, my soapstone for example, will descend from any point in precisely the same time.

Removing the fire-board from the front of the try-works, the bare masonry of that side is exposed, penetrated by the two iron mouths of the furnaces, directly underneath the pots. These mouths are fitted with heavy doors of iron. The intense heat of the fire is prevented from communicating itself to the deck, by means of a shallow reservoir extending under the entire inclosed

Boiling oil on* John R. Manta, *1925. *"Books" of blubber are placed into the large iron "trypots," where the cooking process known as "trying out" will render the blubber into the valuable oil.*

Skimming scraps, *John R. Manta, 1925. After the scraps are skimmed from the top of the boiling oil, the clean oil will be ladled into the cooler and piped to tanks in the hold.*

Boys to men, 1903. *Mincing blubber aboard the* Charles W. Morgan.

William R. Hegarty Collection

William R. Hegarty Collection

surface of the works. By a tunnel inserted at the rear, this reservoir is kept replenished with water as fast as it evaporates. There are no external chimneys; they open direct from the rear wall. And here let us go back for a moment.

It was about nine o'clock at night that the Pequod's try-works were first started on this present voyage. It belonged to Stubb to oversee the business.

"All ready there? Off hatch, then, and start her. You cook, fire the works." This was an easy thing, for the carpenter had been thrusting his shavings into the furnace throughout the passage. Here be it said that in a whaling voyage the first fire in the try-works has to be fed for a time with wood. After that no wood is used, except as a means of quick ignition to the staple fuel. In a word, after being tried out, the crisp, shrivelled blubber, now called scraps or fritters, still contains considerable of its unctuous properties. These fritters feed the flames.

By midnight the works were in full operation. We were clear from the carcase; sail had been made; the wind was freshening; the wild ocean darkness was intense. But that darkness was licked up by the fierce flames, which at intervals forked forth from the sooty flues, and illuminated every lofty rope in the rigging, as with the famed Greek fire. The burning ship drove on, as if remorselessly commissioned to some vengeful deed.

The hatch, removed from the top of the works, now afforded a wide hearth in front of them. Standing on

Trying out. — *Davis:* Nimrod of the Sea.

Feeding the fire. — *William R. Hegarty Collection.*

this were the Tartarean shapes of the pagan harpooneers, always the whale-ship's stokers. With huge pronged poles they pitched hissing masses of blubber into the scalding pots, or stirred up the fires beneath, till the snaky flames darted, curling, out of the doors to catch them by the feet. The smoke rolled away in sullen heaps. To every pitch of the ship there was a pitch of the boiling oil, which seemed all eagerness to leap into their faces. Opposite the mouth of the works, on the further side of the wide wooden hearth, was the windlass. This served for a sea-sofa. Here lounged the watch, when not

Feeding the fire on John R. Manta, 1925. *At first, the tryworks are fueled with wood, then boiled-out pieces of blubber keep the fires going. In this way, the whale supplies the fuel to try out its own blubber.*

William R. Hegarty Collection

159

otherwise employed, looking into the red heat of the fire, till their eyes felt scorched in their heads. Their tawny features, now all begrimed with smoke and sweat, their matted beards, and the contrasting barbaric brilliancy of their teeth, all these were strangely revealed in the capricious emblazonings of the works. As they narrated to each other their unholy adventures, their tales of terror told in words of mirth; as their uncivilized laughter forked upwards out of them, like the flames from the furnace; as to and fro, in their front, the harpooneers wildly gesticulated with their huge pronged forks and dippers; as the wind howled on, and the sea leaped, and the ship groaned and dived, and yet steadfastly shot her red hell further and further into the blackness of the sea and the night, and scornfully champed the white bone in her mouth, and viciously spat round her on all sides; then the rushing Pequod, freighted with savages, and laden with fire, and burning a corpse, and plunging into that blackness of darkness, seemed the material counterpart of her monomaniac commander's soul.

Bark Helen Mar *trying out along the Northwest Coast, 1887.* — *New Bedford Free Public Library.*

Trying out on the* Charles W. Morgan, *circa 1922. *"Like a plethoric burning martyr, or a self-consuming misanthrope, once ignited, the whale supplies his own fuel and burns by his own body. Would that he consumed his own smoke! for his smoke is horrible to inhale, and inhale it you must, and not only that, but you must live in it for the time. It has an unspeakable, wild, Hindoo odor about it, such as may lurk in the vicinity of funereal pyres. It smells like the left wing of the day of judgment; it is an argument for the pit."* — Chapter 96. From "Down to the Sea in Ships."

New Bedford Whaling Museum

Chapter XCVIII

Stowing Down and Clearing Up

Whaleboat and tryworks from *Lucy Ann* of Wilmington.
— *From journal of John F. Martin. Kendall Collection, NBWM.*

Already has it been related how the great leviathan is afar off descried from the mast-head; how he is chased over the watery moors, and slaughtered in the valleys of the deep; how he is then towed alongside and beheaded; and how his great padded surtout becomes the property of his executioner; how, in due time, he is condemned to the pots, and, like Shadrach, Meshach, and Abednego, his spermaceti, oil, and bone pass unscathed through the fire;—but now it remains to conclude the last chapter of this part of the description by rehearsing—singing, if I may—the romantic proceeding of decanting off his oil into the casks and striking them down into the hold, where once again leviathan returns to his native profundities, sliding along beneath the surface as before; but, alas! never more to rise and blow.

While still warm, the oil, like hot punch, is received into the six-barrel casks; and while, perhaps, the ship is pitching and rolling this way and that in the midnight sea, the enormous casks are slewed round and headed over, end for end, and sometimes perilously scoot across the slippery deck, like so many land slides, till at last man-handled and stayed in their course; and all round the hoops, rap, rap, go as many hammers as can play upon them, for now, *ex officio*, every sailor is a cooper.

At length, when the last pint is casked, and all is cool, then the great hatchways are unsealed, the bowels of the ship are thrown open, and down go the casks to their final rest in the sea. This done, the hatches are replaced, and hermetically closed, like a closet walled up.

"The Cachalot Trying-out at Night," circa 1900. *"A South Sea Sperm Whaler, which in a voyage of four years perhaps, after completely filling her hold with oil, does not consume fifty days in the business of boiling out; and in the state that it is casked, the oil is nearly scentless."* — Chapter 92. *Oil painting by William E. Norton, based on "The Cruise of the Cachalot" by Frank T. Bullen. Companion to painting on page 162.*

Kendall Collection, NBWM

In the sperm fishery, this is perhaps one of the most remarkable incidents in all the business of whaling. One day the planks stream with freshets of blood and oil; on the sacred quarter-deck enormous masses of the whale's head are profanely piled; great rusty casks lie about, as in a brewery yard; the smoke from the try-works has besooted all the bulwarks; the mariners go about suffused with unctuousness; the entire ship seems great leviathan himself; while on all hands the din is deafening.

But a day or two after, you look about you, and prick your ears in this self-same ship; and were it not for the tell-tale boats and try-works, you would all but swear you trod some silent merchant vessel, with a most scrupulously neat commander. The unmanufactured sperm oil possesses a singularly cleansing virtue. This is the reason why the decks never look so white as just after what they call an affair of oil. Besides, from the ashes of the burned scraps of the whale, a potent lye is readily made; and whenever any adhesiveness from the back of the whale remains clinging to the side, that lye quickly exterminates it. Hands go diligently along the bulwarks, and with buckets of water and rags restore them to their

Stowing down oil on the **Charles W. Morgan,** *circa 1912.* — *Nelson F. Wood photograph. New Bedford Whaling Museum.*

full tidiness. The soot is brushed from the lower rigging. All the numerous implements which have been in use are likewise faithfully cleansed and put away. The great hatch is scrubbed and placed upon the try-works, completely

"The **Cachalot** *cutting-in," circa 1900.* — *Oil painting by William E. Norton, based on "The Cruise of the Cachalot" by Frank T. Bullen. Companion to painting on page 161.*

Kendall Collection, NBWM

hiding the pots; every cask is out of sight; all tackles are coiled in unseen nooks; and when by the combined and simultaneous industry of almost the entire ship's company, the whole of this conscientious duty is at last concluded, then the crew themselves proceed to their own ablutions; shift themselves from top to toe; and finally issue to the immaculate deck, fresh and all aglow, as bridegrooms new-leaped from out the daintiest Holland.

Now, with elated step, they pace the planks in twos and threes, and humorously discourse of parlors, sofas, carpets, and fine cambrics; propose to mat the deck; think of having hangings to the top; object not to taking tea by moonlight on the piazza of the forecastle. To hint to such musked mariners of oil, and bone, and blubber, were little short of audacity. They know not the thing you distantly allude to. Away, and bring us napkins!

But mark: aloft there, at the three mast heads, stand three men intent on spying out more whales, which, if caught, infallibly will again soil the old oaken furniture, and drop at least one small grease-spot somewhere. Yes; and many is the time, when, after the severest uninterrupted labors, which know no night; continuing straight through for ninety-six hours; when from the boat, where they have swelled their wrists with all day rowing on the Line,—they only step to the deck to carry vast chains, and heave the heavy windlass, and cut and slash, yea, and in their very sweatings to be smoked and burned anew by the combined fires of the equatorial sun and the equatorial

"Cutting in a sperm whale," 1845. — Journal painting by John Francis Akin aboard ship Virginia of New Bedford. Kendall Collection, NBWM.

try-works; when, on the heel of all this, they have finally bestirred themselves to cleanse the ship, and make a spotless dairy room of it; many is the time the poor fellows, just buttoning the necks of their clean frocks, are startled by the cry of "There she blows!" and away they fly to fight another whale, and go through the whole weary thing again. Oh! my friends, but this is man-killing! Yet this is life. For hardly have we mortals by long toilings extracted from the world's vast bulk its small but valuable sperm; and then, with weary patience, cleansed ourselves from its defilements, and learned to live here in clean tabernacles of the soul; hardly is this done, when—*There she blows!*—the ghost is spouted up, and away we sail to fight some other world, and go through young life's old routine again.

Ship Virginia *chasing whales in the Pacific, 1845.* — Journal painting by John Francis Akin aboard ship Virginia of New Bedford.

Kendall Collection, NBWM

Chapter C

Leg and Arm. The Pequod, of Nantucket, meets the Samuel Enderby, of London

Beautifully-crafted cover of the journal kept by William Mott of the ship *Willliam* of London, 1796–99.
— *Artwork by Thomas Wetling. Kendall Collection, NBWM.*

"Ship, ahoy! Hast seen the White Whale?"

So cried Ahab, once more hailing a ship showing English colors, bearing down under the stern. Trumpet to mouth, the old man was standing in his hoisted quarter-boat, his ivory leg plainly revealed to the stranger captain, who was carelessly reclining in his own boat's bow. He was a darkly-tanned, burly, good-natured, fine-looking man, of sixty or thereabouts, dressed in a spacious roundabout, that hung round him in festoons of blue pilot-cloth; and one empty arm of this jacket streamed behind him like the broidered arm of a huzzar's surcoat.

"Hast seen the White Whale?"

"See you this?" and withdrawing it from the fold that had hidden it, he held up a white arm of sperm whale bone, terminating in a wooden head like a mallet.

"Man my boat!" cried Ahab, impetuously, and tossing about the oars near him—"Stand by to lower!"

In less than a minute, without quitting his little craft, he and his crew were dropped to the water, and were soon alongside of the stranger. But here a curious difficulty presented itself. In the excitement of the moment, Ahab had forgotten that since the loss of his leg he had never once stepped on board of any vessel at sea but his own, and then it was always by an ingenious and very handy mechanical contrivance peculiar to the Pequod, and a thing not to be rigged and shipped in any other vessel at a moment's warning.

As good luck would have it, they had had a whale alongside a day or two previous, and the great tackles were still aloft, and the massive curved blubber-hook, now clean and dry, was still attached to the end. This was quickly lowered to Ahab, who at once comprehending it all, slid his solitary thigh into the curve of the hook (it was like sitting in the fluke of an anchor, or the crotch of an apple tree), and then giving the word, held himself fast, and at the same time also helped to hoist

"Ships Atlantic and William cruising for Whale," 1797. British whaleships in the company of Samuel Enderby & Sons of London were the first to round Cape Horn and pursue sperm whaling in the Pacific. The ships traveled up the South American coast toward Peru, then northeast toward the Galapagos Islands where great herds of sperm whales had been seen by Enderby's exploration ships. By the 1840s, the Pacific Ocean just west of the Galapagos would become one of the greatest of all sperm whaling grounds. — *Source: Dakin. Watercolor by Thomas Wetling, from journal of the ship William of London.*

Kendall Collection, NBWM

his own weight, by pulling hand-over-hand upon one of the running parts of the tackle. Soon he was carefully swung inside the high bulwarks, and gently landed upon the capstan head. With his ivory arm frankly thrust forth in welcome, the other captain advanced, and Ahab, putting out his ivory leg, and crossing the ivory arm (like two sword-fish blades) cried out in his walrus way, "Aye, aye, hearty! let us shake bones together!—an arm and a leg!—an arm that never can shrink, d'ye see; and a leg that never can run. Where did'st thou see the White Whale?—how long ago?"

"The White Whale," said the Englishman, pointing his ivory arm towards the East, and taking a rueful sight along it, as if it had been a telescope; "There I saw him, on the Line, last season."

"And he took that arm off, did he?" asked Ahab, now sliding down from the capstan, and resting on the Englishman's shoulder, as he did so.

"Aye, he was the cause of it, at least; and that leg, too?"

"Spin me the yarn," said Ahab; "how was it?"

"It was the first time in my life that I ever cruised on the Line," began the Englishman. "I was ignorant of the White Whale at that time. Well, one day we lowered for a pod of four or five whales, and my boat fastened to one of them; a regular circus horse he was, too, that went milling and milling round so, that my boat's crew could only trim dish, by sitting all their sterns on the outer gunwale. Presently up breaches from the bottom of the sea a bouncing great whale, with a milky-white head and hump, all crows' feet and wrinkles."

"It was he, it was he!" cried Ahab, suddenly letting out his suspended breath.

"How it was exactly," continued the one-armed commander, "I do not know; but in biting the line, it got foul of this teeth, caught there somehow; but we didn't know it then; so that when we afterwards pulled on the line, bounce we came plump on to his hump! Seeing how matters stood, and what a noble great whale it was—the noblest and biggest I ever saw, sir, in my life—I resolved to capture him, spite of the boiling rage he seemed to be in. I jumped into my first mate's boat; and snatching the first harpoon, let this old great-grandfather have it. But, Lord, look you, sir—hearts and souls alive, man—the next instant, in a jiff, I was blind as a bat—both eyes out—all befogged and bedeadened with black foam—the whale's tail looming straight up out of it, perpendicular in the air, like a marble steeple. As I was groping after the second iron, to toss it overboard—down comes the tail like a Lima tower, cutting my boat in two, leaving each half in splinters; and, flukes first, the white hump backed through the wreck, as though it was all chips. To escape his terrible flailings, I seized hold of my harpoon-pole sticking in him, and for a moment clung to that like a sucking fish. But a combing sea dashed me off, and at the same instant, the fish, taking one good dart forwards, went down like a flash; and the barb of that cursed second iron towing

"A View of the Boats going after the Whale," circa 1798. In his petition to the Lords of the Admiralty to break the East India monopoly on commerce along South Pacific ocean routes serving the East, Enderby wrote: "We appear to be the only adventurers willing to risk their property at such a great distance for the exploring of a Fishery… If [our ship] is successful a large Branch of the Fishery will be carried on in those seas." — Source: Dakin, Whalemen Adventureres. *Watercolor by Thomas Wetling, from the journal of the ship* William of London.

Kendall Collection, NBWM

along near me caught me here" (clapping his hand just below his shoulder); "yes, caught me just here, I say, and bore me down to Hell's flames, I was thinking; when, when, all of a sudden, the barb ript its way along the flesh clear along the whole length of my arm—came out nigh my wrist, and up I floated;—and that gentleman will tell you the rest (by the way, Captain—Dr. Bunger, ship's surgeon: Bunger, my lad—the captain)."

"It was a shocking bad wound," began the whale-surgeon; "the truth was, sir, it was an ugly gaping wound as surgeon ever saw; more than two feet and several inches long. In short, it grew black; I knew what was threatened, and off it came. But I had no hand in shipping that ivory arm there; that is the captain's work, not mine."

"What became of the White Whale?"

"Oh!" cried the one-armed captain, "Oh, yes! Well; after he sounded, we didn't see him again for some time; in fact, as I before hinted, I didn't then know what whale it was that had served me such a trick, till some time afterwards, when coming back to the Line, we heard about Moby Dick—as some call him—and then I knew it was he."

"Did'st thou cross his wake again?"

"Twice."

"But could not fasten?"

"No more White Whales for me; I've lowered for him once, and that has satisfied me. There would be great glory in killing him, I know that; and there is a ship-load of precious sperm in him, but, hark ye, he's best let alone; don't you think so, Captain?"—glancing at the ivory leg.

"He is. But he will still be hunted, for all that. What is best let alone, that accursed thing is not always what least allures. He's all a magnet! How long since thou saw'st him last? Which way heading?"

"Good God!" cried the English Captain, to whom the question was put. "What's the matter? He was heading east, I think.—Is your Captain crazy?" whispering to Fedallah.

But Fedallah, putting a finger on his lip, slid over the bulwarks to take the boat's steering oar, and Ahab, swinging the cutting-tackle towards him, commanded the ship's sailors to stand by to lower.

In a moment he was standing in the boat's stern, and the Manilla men were springing to their oars. In vain the English Captain hailed him. With back to the stranger ship, and face set like a flint to his own, Ahab stood upright till alongside of the Pequod.

"**The South Sea Whale Fishery,**" *circa 1835.* Following Enderby in 1785, the British continued to hunt the South Pacific, and the Americans were right on their heels. After recovering from the War of Independence (1783), the devasted whaling industry got a boost when the Minister to England, John Adams, brokered treaties throughout Europe that included the purchase of American spermacetti oil. In 1791, six ships from Nantucket and one from New Bedford were the first to set sail for the Pacific. — *Source: Starbuck:* History…Whale Fishery. *Oil painting by John Ward, London.*

Kendall Collection, NBWM

Chapter ci

The Decanter

Ere the English ship fades from sight, be it set down here, that she hailed from London, and was named after the late Samuel Enderby, merchant of that city, the original of the famous whaling house of Enderby and Sons. How long, prior to the year of our Lord 1775, this great whaling house was in existence, my numerous fish-documents do not make plain; but in that year (1775) it fitted out the first English ships that ever regularly hunted the Sperm Whale; though for some score of years previous (ever since 1726) our valiant Coffins and Maceys of Nantucket and the Vineyard had in large fleets pursued that Leviathan, but only in the North and South Atlantic: not elsewhere. Be it distinctly recorded here, that the Nantucketers were the first among mankind to harpoon with civilized steel the great Sperm Whale; and that for half a century they were the only people of the whole globe who so harpooned him.

In 1778, a fine ship, the Amelia, fitted out for the express purpose, and at the sole charge of the vigorous Enderbys, boldly rounded Cape Horn, and was the first among the nations to lower a whale-boat of any sort in the great South Sea. The voyage was a skilful and lucky one; and returning to her berth with her hold full of the precious sperm, the Amelia's example was soon followed by other ships, English and American, and thus the vast Sperm Whale grounds of the Pacific were thrown open. But this is not all. In 1819, the same house fitted out a discovery whale ship, to go on a tasting cruise to the remote waters of Japan. That ship—well called the "Syren"—made a noble experimental cruise; and it was thus that the great Japanese Whaling Ground first became generally known. The Syren in this famous voyage was commanded by a Captain Coffin, a Nantucketer.

Whaleship Adam of London, circa 1817. *This watercolor by an anonymous British seaman is the same scene as the earliest pictorial work on whale ivory yet discovered—a large engraved sperm whale tooth, inscribed: "This is the tooth of a sperm whale that was caught near the Galapagos Island by the crew of the ship Adam, and made 100 barrels of oil in the year 1817." The watercolor was probably done at around the same time, possibly by the same hand. The ship was built at Duxbury, Massachusetts in 1794, and completed several Atlantic Ocean whaling voyages as the Renown of Nantucket. Like many ships, it was boarded and taken as an admiralty prize during the War of 1812. The British renamed her Adam and she made at least four South Seas voyages (1815–25) under London registration.* — Source: Frank, Melville's…Gallery.

"Ship William on Her Passage to Cape Horn, and on the Coast of Peru," circa 1797. *Melville's information in this chapter is inaccurate: Samuel Enderby sent his ship Emilia (not "Amelia") around Cape Horn in 1785 (not 1775). Enderby knew the Americans would soon be sending ships. He had received reports from an American merchant returning from a trading voyage to China who "had seen more Spermaceti Whales about the Straights of Sunda and the Island of Java than he had ever seen before, so much so that he could have filled a ship of 300 tons in 3 months." The American wanted Enderby to sponsor a voyage, but Enderby was more interested in putting his own ships in service. He was quick to act. "We have 2 ships of 300 tons each of which we are beginning to fit for the Southern Whale Fishery, to sail in March…." The first American whaler to return from the South Pacific with oil was the ship Rebecca of New Bedford in 1793, with 750 barrels of sperm whale oil.* — Sources: Davit, Starbuck. Watercolor by Thomas Wetling, from the journal of the ship William of London.

Kendall Collection, NBWM

Chapter CII

A Bower in the Arsacides

Skeleton of a blue whale on exhibit at New Bedford Whaling Museum.

I confess, that since Jonah, few whalemen have penetrated very far beneath the skin of the adult whale; nevertheless, I have been blessed with an opportunity to dissect him in miniature. In a ship I belonged to, a small cub sperm whale was once bodily hoisted to the deck for his poke or bag, to make sheaths for the barbs of the harpoons, and for the heads of the lances. And as for my exact knowledge of the bones of the leviathan in their gigantic, full grown development, for that rare knowledge I am indebted to my late royal friend Tranquo, king of Tranque, one of the Arsacides.

Among many other fine qualities, my royal friend Tranquo, being gifted with a devout love for all matters of barbaric vertu, had brought together in Pupella whatever rare things the more ingenious of his people could invent; chiefly carved woods of wonderful devices, chiselled shells, inlaid spears, costly paddles, aromatic canoes; and all these distributed among whatever natural wonders, the wonder-freighted, tribute-rendering waves had cast upon his shores.

Chief among these latter was a great sperm whale, which, after an unusually long raging gale, had been found dead and stranded, with his head against a cocoa-nut tree, whose plumage-like, tufted droopings seemed his verdant jet. When the vast body had at last been stripped of its fathom-deep enfoldings, and the bones become dust dry in the sun, then the skeleton was carefully transported up the Pupella glen, where a grand temple of lordly palms now sheltered it.

Now, amid the green, life-restless loom of that Arsacidean wood, the great, white, worshipped skeleton lay lounging—a gigantic idler! Yet, as the ever-woven verdant warp and woof intermixed and hummed around him, the mighty idler seemed the cunning weaver; himself all woven over with the vines; every month assuming greener, fresher verdure; but himself a skeleton. Life folded Death; Death trellised Life; the grim god wived with youthful Life, and begat him curly-headed glories.

The skeleton dimensions I shall now proceed to set down are copied verbatim from my right arm, where I had them tattooed; as in my wild wanderings at that period, there was no other secure way of preserving such valuable statistics.

"The Spermacetti Whale," 1837. *The length of the adult male is 49-59 feet and weighs 45-70 tons. The female is smaller at 35–40 feet and 15–20 tons. The world population is estimated at between 100,000–200,000.* — *Drawing by Stewart, engraved by W. Lizars, colored lithograph from Hamilton:* The Natural History....

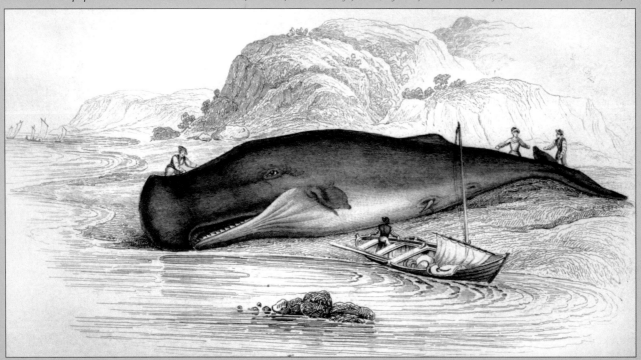

New Bedford Whaling Museum

Chapter CIII

Measurement of the Whale's Skeleton

"Skeleton of the Great Northern Rorqual," 1837. — *From Hamilton.*

In the first place, I wish to lay before you a particular, plain statement, touching the living bulk of this leviathan, whose skeleton we are briefly to exhibit. Such a statement may prove useful here.

According to a careful calculation I have made, and which I partly base upon Captain Scoresby's estimate, of seventy tons for the largest sized Greenland whale of sixty feet in length; according to my careful calculation, I say, a sperm whale of the largest magnitude, between eighty-five and ninety feet in length, and something less than forty feet in its fullest circumference, such a whale will weigh at least ninety tons; so that reckoning thirteen men to a ton, he would considerably outweigh the combined population of a whole village of one thousand one hundred inhabitants.

There are forty and odd vertebrae in all, which in the skeleton are not locked together. They mostly lie like the great knobbed blocks on a Gothic spire, forming solid courses of heavy masonry. The largest, a middle one, is in width something less than three feet, and in depth more than four. The smallest, where the spine tapers away into the tail, is only two inches in width, and looks something like a white billiard-ball. I was told that there were still smaller ones, but they had been lost by some little cannibal urchins, the priest's children, who had stolen them to play marbles with. Thus we see how that the spine of even the hugest of living things tapers off at last into simple child's play.

"The Lesser Rorqual," 1837. The Minke whale (also known as lesser rorqual, lesser finback and pikehead) is the smallest of the rorqual or baleen whales. The male can grow to 32 feet and the female 36 feet and weigh 10 tons. The smallest of the seven great whales, its size made it uneconomical to harvest commercially. — *Colored lithograph from Hamilton.*

"Northern Rorqual," 1837. The blue whale (also called sulphur bottom, blue rorqual or great blue whale) is the largest animal on earth. Most adults are 75 to 90 feet long and weigh 100-120 tons. The longest recorded blue whale was over 109 feet, and the heaviest weighed about 190 tons. Their population is about 4,500. — *Colored lithograph from Hamilton.*

New Bedford Whaling Museum

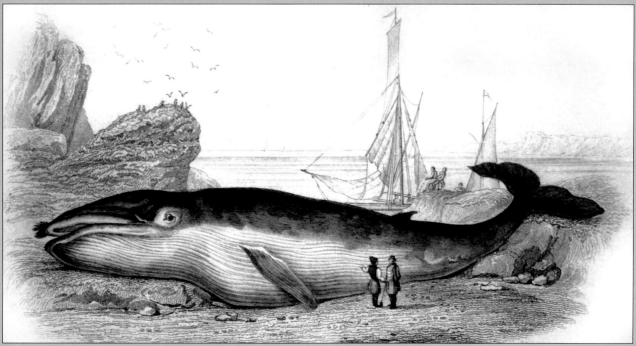

New Bedford Whaling Museum

169

Chapter CV

Does the Whale's Magnitude Diminish?—Will he Perish?

Logbook art by John C. Scales aboard bark *Pearl* of New London, 1852-54. — *Kendall Collection, NBWM*

But still another inquiry remains; one often agitated by the more recondite Nantucketers. Whether owing to the almost omniscient look-outs at the mastheads of the whale-ships, now penetrating even through Behring's straits, and into the remotest secret drawers and lockers of the world; and the thousand harpoons and lances darted along all continental coasts; the moot point is, whether leviathan can long endure so wide a chase, and so remorseless a havoc; whether he must not at last be exterminated from the waters, and the last whale, like the last man, smoke his last pipe, and then himself evaporate in the final puff.

Comparing the humped herds of whales with the humped herds of buffalo, which, not forty years ago, overspread by tens of thousands the prairies of Illinois and Missouri, and shook their iron manes and scowled with their thunder-clotted brows upon the sites of populous river-capitals, where now the polite broker sells you land at a dollar an inch; in such a comparison an irresistible argument would seem furnished, to show that the hunted whale cannot now escape speedy extinction.

We account the whale immortal in his species, however perishable in his individuality. He swam the seas before the continents broke water; he once swam over the site of the Tuileries, and Windsor Castle, and the Kremlin. In Noah's flood, he despised Noah's Ark; and if ever the world is to be again flooded, like the Netherlands, to kill off its rats, then the eternal whale will still survive, and rearing upon the topmost crest of the equatorial flood, spout his frothed defiance to the skies.

The Spermacetti Whale, 1837. *"Since I have undertaken to manhandle this Leviathan, it behoves me to approve myself omnisciently exhaustive in the enterprise… To produce a mighty book, you must choose a mighty theme. No great and enduring volume can ever be written on the flea, though many there be who have tried it."* — *Chapter 104. Colored lithograph from Hamilton.*

New Bedford Whaling Museum

Chapter CVI, CVII, CVIII

Ahab's Leg · The Carpenter · Ahab and the Carpenter

The precipitating manner in which Captain Ahab had quitted the Samuel Enderby of London, had not been unattended with some small violence to his own person. He had lighted with such energy upon a thwart of his boat that his ivory leg had received a half-splintering shock. And when after gaining his own deck, and his own pivot-hole there, he so vehemently wheeled round with an urgent command to the steersman (it was, as ever, something about his not steering inflexibly enough); then, the already shaken ivory received such an additional twist and wrench, that though it still remained entire, and to all appearances lusty, yet Ahab did not deem it entirely trustworthy.

He took plain practical procedures;-he called the carpenter. And when that functionary appeared before him, he bade him without delay set about making a new leg, and directed the mates to see him supplied with all the studs and joists of jaw-ivory (sperm whale) which had thus far been accumulated on the voyage, in order that a careful selection of the stoutest, clearest-grained stuff might be secured. This done, the carpenter received orders to have the leg completed that night; and to provide all the fittings for it, independent of those pertaining to the distrusted one in use.

Chapter CVII

Like all sea-going ship carpenters, and more especially those belonging to whaling vessels, he was alike experienced in numerous trades and callings collateral to his own.

A lost land-bird of strange plumage strays on board, and is made a captive: out of clean shaved rods of right-whale bone, and cross-beams of sperm whale ivory, the carpenter makes a pagoda-looking cage for it. An oarsman sprains his wrist: the carpenter concocts a soothing lotion. Stubb longed for vermillion stars to be painted upon the blade of his every oar; screwing each oar in his big vice of wood, the carpenter symmetrically supplies the constellation. A sailor takes a fancy to wear shark-bone ear-rings: the carpenter drills his ears. Another has the toothache: the carpenter out pincers, and clapping one hand upon his bench bids him be seated there.

Thus, this carpenter was prepared at all points, and alike indifferent and without respect in all. Teeth he accounted bits of ivory; heads he deemed but top-blocks; men themselves he lightly held for capstans. He

Bird cage made of whalebone.
—Anonymous. New Bedford Whaling Museum.

was like one of those unreasoning but still highly useful, multum in parvo, Sheffield contrivances, assuming the exterior—though a little swelled—of a common pocket knife; but containing, not only blades of various sizes, but also screw-drivers, cork-screws, tweezers, awls, pens, rulers, nail-filers, counter-sinkers. So, if his superiors wanted to use the carpenter for a screw-driver, all they had to do was to open that part of him, and the screw was fast: or if for tweezers, take him up by the legs, and there they were.

Chapter CVIII

The Deck—First Night Watch

(Carpenter standing before his vice-bench, and by the light of two lanterns busily filing the ivory joist for the leg, which joist is firmly fixed in the vice. Slabs of ivory, leather straps, pads, screws, and various tools of all sorts lying about the bench. Forward, the red flame of the forge is seen, where the blacksmith is at work.)

"Well, manmaker!"

"Just in time, Sir. If the captain pleases, I will now mark the length. Let me measure, Sir."

"Measured for a leg! good. Well, it's not the first time. About it! Look ye, carpenter, I dare say thou callest thyself a right good workmanlike workman, eh! Well, then, will it speak thoroughly well for thy work, if, when I come to mount this leg thou makest, I shall nevertheless feel another leg in the same identical place with it; that is, carpenter, my old lost leg; the flesh and blood one, I mean. Canst thou not drive that old Adam away?"

"I have heard something curious on that score, Sir; how that a dismasted man never entirely loses the feeling of his old spar, but it will be still pricking him at times. May I humbly ask if it be really so, Sir?"

"It is, man. Look, put thy live leg here in the place where mine once was; so, now, here is only one distinct leg to the eye, yet two to the soul. Where thou feelest tingling life; there, exactly there, there to a hair, do I. Is't a riddle?"

"I should humbly call it a poser, Sir."

"How long before this leg is done?"

"Perhaps an hour, Sir."

Chapter CIX

Ahab and Starbuck in the Cabin

According to usage they were pumping the ship next morning; and lo! no inconsiderable oil came up with the water; the casks below must have sprung a bad leak. Much concern was shown; and Starbuck went down into the cabin to report this unfavorable affair.[1]

Now, from the South and West the Pequod was drawing nigh to Formosa and the Bashee Isles, between which lies one of the tropical outlets from the China waters into the Pacific. And so Starbuck found Ahab with a general chart of the oriental archipelagoes spread before him; and another separate one representing the long eastern coasts of the Japanese islands—Niphon, Matsmai, and Sikoke.

"The oil in the hold is leaking, sir. We must up Burtons and break out."

"Up Burtons and break out? Now that we are nearing Japan; heave-to here for a week to tinker a parcel of old hoops?"

"Either do that, sir, or waste in one day more oil than we may make good in a year. What we come twenty thousand miles to get is worth saving, sir."

"So it is, so it is; if we get it."

"I was speaking of the oil in the hold, sir."

"And I was not speaking or thinking of that at all. Begone! Let it leak! I'm all aleak myself. Yet I don't stop to plug my leak; for who can find it in the deep-loaded hull; or how hope to plug it, even if found, in this life's howling gale? Starbuck! I'll not have the Burtons hoisted."

"What will the owners say, sir?"

"Let the owners stand on Nantucket beach and outyell the Typhoons. What cares Ahab? Owners, own-

Crosscut view of the bark *Alice Knowles* showing location of oil and provision casks. — *National Archives.*

ers? Thou art always prating to me, Starbuck, about those miserly owners, as if the owners were my conscience. But look ye, the only real owner of anything is its commander; and hark ye, my conscience is in this ship's keel.—On deck!"

Ahab seized a loaded musket from the rack (forming part of most South-Sea-men's cabin furniture), and pointing it towards Starbuck, exclaimed: "There is one God that is Lord over the earth, and one Captain that is lord over the Pequod.—On deck!"

For an instant in the flashing eyes of the mate, and his fiery cheeks, you would have almost thought that he had really received the blaze of the levelled tube. But, mastering his emotion, he half calmly rose, and as he quitted the cabin, paused for an instant and said: "Thou hast outraged, not insulted me, Sir; but for that I ask thee not to beware of Starbuck; thou wouldst but laugh; but let Ahab beware of Ahab; beware of thyself, old man."

"He waxes brave, but nevertheless obeys; most careful bravery that!" murmured Ahab, as Starbuck disappeared. "What's that he said—Ahab beware of Ahab—there's something there!" Then unconsciously using the musket for a staff, with an iron brow he paced to and fro in the little cabin; but presently the thick plaits of his forehead relaxed, and returning the gun to the rack, he went to the deck.

"Thou art but too good a fellow, Starbuck," he said lowly to the mate; then raising his voice to the crew: "Furl the t'gallant-sails and close-reef the top-sails, fore and aft; back the main-yard; up Burtons, and break out in the main-hold."

It were perhaps vain to surmise exactly why it was, that as respecting Starbuck, Ahab thus acted. It may have been a flash of honesty in him; or mere prudential policy which, under the circumstance, imperiously forbade the slightest symptom of open disaffection, however transient, in the important chief officer of his ship. However it was, his orders were executed; and the Burtons were hoisted.

"Booming along," near the island of Fernando Noronha, 1846. *The ship* William and Eliza *of New Bedford, captained by William H. Whitfield, makes high speed in strong winds, leaving no time to "speak," or talk to, passing ships.* — *Watercolor from the journal of Francis Marion Shaw.*

[1] In sperm-whalemen with any considerable quantity of oil on board, it is a regular semi-weekly duty to conduct a hose into the hold, and drench the casks with sea-water; which afterwards, is removed by the ship's pumps. By the changed character of the withdrawn water, the mariners readily detect any serious leakage in the precious cargo.

Chapter cx

Queequeg in his Coffin

Upon searching, it was found that the casks last struck into the hold were perfectly sound, and that the leak must be further off. So, it being calm weather, they broke out deeper and deeper, disturbing the slumbers of the huge ground-tier butts; and from that black midnight sending those gigantic moles into the daylight above.

Now, at this time it was that my poor pagan companion, and fast bosom-friend, Queequeg, was seized with a fever, which brought him nigh to his endless end.

Poor Queequeg! When the ship was about half disembowelled, you should have stooped over the hatchway, and peered down upon him there; where, stripped to his woollen drawers, the tattooed savage was crawling about amid that dampness and slime, like a green spotted lizard at the bottom of a well. And a well, or an ice-house, it somehow proved to him, poor pagan; where, strange to say, for all the heat of his sweatings, he caught a terrible chill which lapsed into a fever; and at last, after some days' suffering, laid him in his hammock, close to the very sill of the door of death.

Not a man of the crew but gave him up; and, as for Queequeg himself, what he thought of his case was forcibly shown by a curious favor he asked. He shuddered at the thought of being buried in his hammock, according to the usual sea-custom, tossed like something vile to the death-devouring sharks. No: he desired a canoe like those of Nantucket, all the more congenial to him, being a whaleman, that like a whale-boat these coffin-canoes were without a keel; though that involved but uncertain steering, and much lee-way adown the dim ages.

The carpenter was at once commanded to do Queequeg's bidding, whatever it might include. There was some heathenish, coffin-colored old lumber aboard, and from these dark planks the coffin was to be made.

"Ah! poor fellow! He'll have to die now," ejaculated the Long Island sailor.

When the last nail was driven, and the lid duly planed and fitted, he lightly shouldered the coffin and went forward with it, inquiring whether they were ready for it yet in that direction. Queequeg, to every one's consternation, commanded that the thing should be instantly brought to him, nor was there any denying him.

Leaning over in his hammock, Queequeg long regarded the coffin with an attentive eye. He then called for his harpoon, had the wooden stock drawn from it, and then had the iron part placed in the coffin along with

Seaman's chest, circa 1870. The sea chest of Manoel E. de Mendonça features a lush scene of New Bedford Harbor painted on the lid. Even the sturdiest of chests couldn't protect the whaleman's clothes from his arduous, oily work. Whalemen quickly found themselves drawing from the ship's slop chest to replace ruined clothing. At the end of the voyage, the whaleman often owed to the ship a good portion of his earnings.
— On exhibit at New Bedford Whaling Museum.

one of the paddles of his boat. All by his own request, also, biscuits were then ranged round the sides within: a flask of fresh water was placed at the head, and a small bag of woody earth scraped up in the hold at the foot; and a piece of sail-cloth being rolled up for a pillow, Queequeg now entreated to be lifted into his final bed, that he might make trial of its comforts, if any it had. He lay without moving a few minutes, then told one to go to his bag and bring out his little god, Yojo. Then crossing his arms on his breast with Yojo between, he called for the coffin lid (hatch he called it) to be placed over him. The head part turned over with a leather hinge, and there lay Queequeg in his coffin with little but his composed countenance in view. "Rarmai" (it will do; it is easy), he murmured at last, and signed to be replaced in his hammock.

But now that he had apparently made every preparation for death; now that his coffin was proved a good fit, Queequeg suddenly rallied; soon there seemed no need of the carpenter's box: and thereupon, when some expressed their delighted surprise, he, in substance, said, that the cause of his sudden convalescence was this;–at a critical moment, he had just recalled a little duty ashore, which he was leaving undone; and therefore had changed his mind about dying; it was Queequeg's conceit, that if a man made up his mind to live, mere sickness could not kill him: nothing but a whale, or a gale, or some violent, ungovernable, unintelligent destroyer of that sort.

With a wild whimsiness, he now used his coffin for a sea-chest; and emptying into it his canvas bag of clothes, set them in order there. Many spare hours he spent, in carving the lid with all manner of grotesque figures and drawings; and it seemed that hereby he was striving, in his rude way, to copy parts of the twisted tattooing on his body.

Chapter CXI

The Pacific

When gliding by the Bashee isles we emerged at last upon the great South Sea; were it not for other things, I could have greeted my dear Pacific with uncounted thanks, for now the long supplication of my youth was answered; that serene ocean rolled eastwards from me a thousand leagues of blue.

There is, one knows not what sweet mystery about this sea, whose gently awful stirrings seem to speak of some hidden soul beneath; for here, millions of mixed shades and shadows, drowned dreams, somnambulisms, reveries; all that we call lives and souls, lie dreaming, dreaming, still; tossing like slumberers in their beds; the ever-rolling waves but made so by their restlessness.

To any meditative Magian rover, this serene Pacific, once beheld, must ever after be the sea of his adoption. It rolls the midmost waters of the world, the Indian ocean and Atlantic being but its arms. The same waves wash the moles of the new-built Californian towns, but yesterday planted by the recentest race of men, and lave the faded but still gorgeous skirts of Asiatic lands, older than

Pacific pit stops, 1845. Had Ahab agreed to "up Burtons and break out," the Pequod's crew may have been treated to a respite similar to these scenes from Russell and Purrington's panorama, "Whaling Voyage Round the World."

At top, friendly natives from the Fiji Islands venture out in fancy canoes and catamarans—their huts and temples nestled between a warm magenta sky and white sandy beach. Below right, an American whaler anchored in Kealakekua Bay near the big island of Hawaii receives friendly Hawaiian greetings; to the left, whalers rest at anchor in Lahaina Habor, Maui. At bottom, a scene from Huahine of the Society Islands showing village huts and missionary settlements, people swimming, a raft of casks being towed to a vessel, and a whaleship at far left nestled in a shady grove of palm trees and friendly inhabitants.

New Bedford Whaling Museum

175

***"Ship* William *on her Passage to Cape Horn,"* 1795.**
Watercolor by Thomas Wetling, from the journal of the ship William *of London. Kendall*

Whaleships Kutusoff *and* Falcon *sperm whaling in the Pacific,* 1846. *Just 8 weeks before Melville set sail on the Acushnet, Benjamin Russell embarked on a 4-year voyage to the Pacific aboard the Kutusoff. Unlike Melville, he stuck it out to the end and returned to New Bedford to repay his debts. This is one of Russell's earliest known ship portraits, painted around the time he and Purrington were creating their panorama.*

Kendall Collection, NBWM

Forbes Collection of the MIT Museum

Abraham; while all between float milky-ways of coral isles, and low-lying, endless, unknown Archipelagoes, and impenetrable Japans. Thus this mysterious, divine Pacific zones the world's whole bulk about; makes all coasts one bay to it; seems the tide-beating heart of earth. Lifted by those eternal swells, you needs must own the seductive god, bowing your head to Pan.

But few thoughts of Pan stirred Ahab's brain, as standing like an iron statue at his accustomed place beside the mizen rigging, with one nostril he unthinkingly snuffed the sugary musk from the Bashee isles (in whose sweet woods mild lovers must be walking), and with the other consciously inhaled the salt breath of the new found sea; that sea in which the hated White Whale must even then be swimming. Launched at length upon these almost final waters, and gliding towards the Japanese cruising-ground, the old man's purpose intensified itself. His firm lips met like the lips of a vice; the Delta of his forehead's veins swelled like overladen brooks; in his very sleep, his ringing cry ran through the vaulted hull, "Stern all! the White Whale spouts thick blood!"

Society Islands, 1845. — *From Russell and Purrington's panorama.*

Bark Canton bound for the Pacific, circa 1915. — *Photograph by Pardon Gifford from the deck of the John R. Manta.*

Private Collection

Melville Voyages

The Acushnet. The white line **(1)** marks the Acushnet's 1842 maiden voyage with Melville aboard. Typical of whalers destined for the Pacific, her route "doubles" Cape Horn. The white line turns black **(2)** after the Marquesas, where Melville jumped ship in July of 1843. The Acushnet continued sailing around the South Pacific before heading to the Northwest Coast. She returned to Fairhaven in 1845 with 850 barrels of sperm oil, 1350 barrels of whale oil, and 13,500 whalebone.

The Pequod. The gray line **(3)** represents the voyage of the Pequod, which, unlike Acushnet, set out to hunt a specific whale. Like the Acushnet, the Pequod sailed east to the Azores, turned south toward the Cape Verde Islands, then into the South Atlantic and southeast to Rio de la Plata, Argentina. There, the Pequod's voyage deviates from that of the Acushnet's and she heads east toward the Cape of Good Hope en route to the Indian and Pacific Oceans.

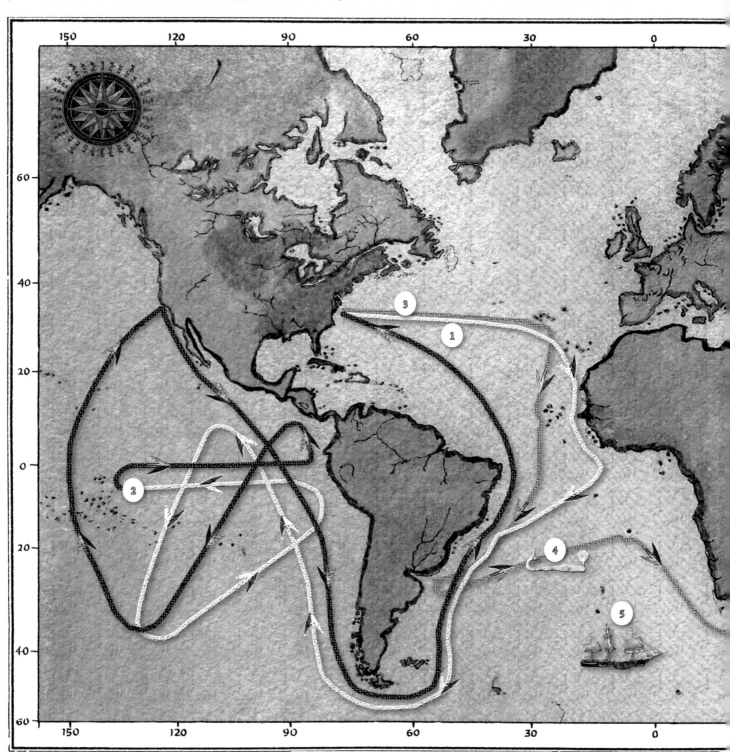

In the mid-Atlantic, her crew sees apparitions of Moby-Dick (4). Not until she's closer to the Cape does she finally lower for whales (5). Near the Crozet Islands, she meets the homeward-bound Albatross *and gams (6) with the* Town-Ho. *In the Indian Ocean, the* Pequod *encounters a giant squid, a school of right whales, and Stubb kills a whale (7). In the Java Sea, she attacks a "grand armada" of whales (8), meets the* Rose-Bud, *the* Enderby *and the* Decanter, *and tries-out oil at night (9).*

The Pequod *then enjoys the rich sperm whaling grounds in the Japan Sea (Inset), and encounters the* Virgin, *the* Bachelor *(10), the* Rachel *(11) and the* Delight *(12) before beginning her chase of Moby-Dick (13 and 14).*

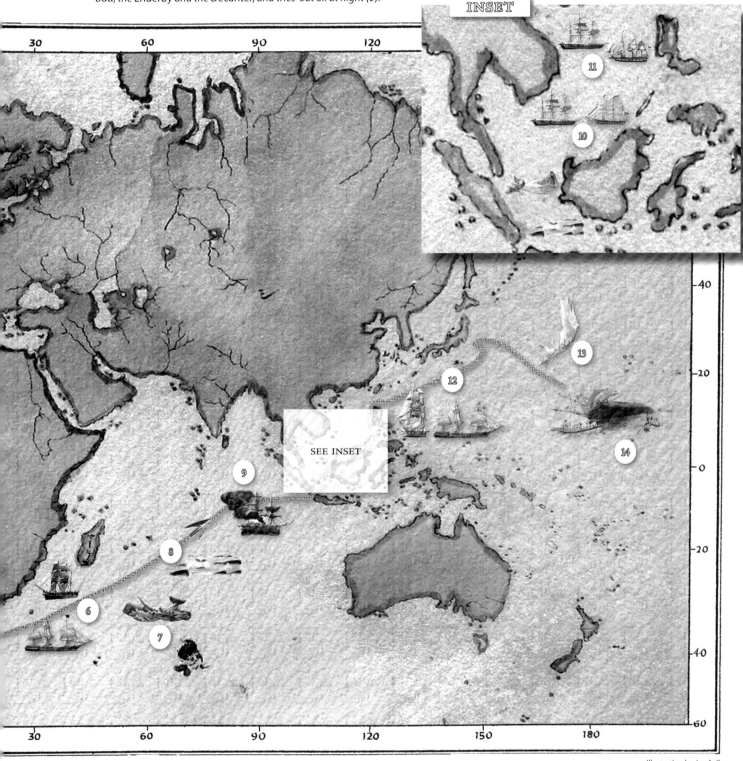

Illustration by Jay Avila

Chapter CXII, CXIII

The Blacksmith · The Forge

Availing himself of the mild, summer-cool weather that now reigned in these latitudes, and in preparation for the peculiarly active pursuits shortly to be anticipated, Perth, the begrimed, blistered old blacksmith, had not removed his portable forge to the hold again, after concluding his contributory work for Ahab's leg, but still retained it on deck, fast lashed to ringbolts by the foremast; being now almost incessantly invoked by the headsmen, and harpooneers, and bowsmen to do some little job for them; altering, or repairing, or new shaping their various weapons and boat furniture. Often he would be surrounded by an eager circle, all waiting to be served; holding boat-spades, pike-heads, harpoons, and lances, and jealously watching his every sooty movement, as he toiled. Nevertheless, this old man's was a patient hammer wielded by a patient arm. No murmur, no impatience, no petulence did come from him. Silent, slow, and solemn; bowing over still further his chronically broken back, he toiled away, as if toil were life itself, and the heavy beating of his hammer the heavy beating of his heart. And so it was.—Most miserable!

Death seems the only desirable sequel for a career like this; but Death is only a launching into the region of the strange Untried; it is but the first salutation to the possibilities of the immense Remote, the Wild, the Watery, the Unshored; therefore, to the death-longing eyes of such men, who still have left in them some interior compunctions against suicide, does the all-contributed and all-receptive ocean alluringly spread forth his whole plain of unimaginable, taking terrors, and wonder-

A whaling-era blacksmith forges away, New Bedford, circa 1880s. — *Joseph G. Tirrell photograph.*

ful, new-life adventures; and from the hearts of infinite Pacifics, the thousand mermaids sing to them—"Come hither, broken-hearted; here is another life without the guilt of intermediate death; here are wonders supernatural, without dying for them. Come hither! bury thyself in a life which, to your now equally abhorred and abhorring, landed world, is more oblivious than death. Come hither! put up thy grave-stone, too, within the churchyard, and come hither, till we marry thee!"

Hearkening to these voices, East and West, by early sun-rise, and by fall of eve, the blacksmith's soul responded, Aye, I come! And so Perth went a-whaling.

CHAPTER CXIII

With matted beard, and swathed in a bristling shark-skin apron, about mid-day, Perth was standing between his forge and anvil, the latter placed upon an iron-wood log, with one hand holding a pike-head in the coals, and with the other at his forge's lungs, when Captain Ahab came along, carrying in his hand a small rusty-looking leathern bag.

"What wert thou making there?"

"Welding an old pike-head, sir; there were seams and dents in it."

"And can'st thou make it all smooth, again, blacksmith, after such hard usage as it had?"

"I think so, sir."

"And I suppose thou can'st smoothe almost any seams and dents; never mind how hard the metal, blacksmith?"

"Aye, sir, I think I can; all seams and dents but one."

Sharpening a lance on the John R. Manta, *1925.* **Tool maintenance was important business aboard ship.**

William R. Hegarty Collection

"Look ye here, then," cried Ahab, passionately advancing, and leaning with both hands on Perth's shoulders; sweeping one hand across his ribbed brows; "Can'st thou smoothe this seam?"

"Oh! that is the one, Sir! Said I not all seams and dents but one?"

"Look ye here!" jingling the leathern bag, as if it were full of gold coins. "I, too, want a harpoon made; one that a thousand yoke of fiends could not part, Perth; something that will stick in a whale like his own fin-bone. There's the stuff, flinging the pouch upon the anvil. Look ye, blacksmith, these are the gathered nail-stubbs of the steel shoes of racing horses."

"Horse-shoe stubs, Sir? Why, Captain Ahab, thou hast here, then, the best and stubbornest stuff we blacksmiths ever work."

"I know it, old man; these stubs will weld together like glue from the melted bones of murderers. Quick! forge me the harpoon. And forge me first, twelve rods for its shank; then wind, and twist, and hammer these twelve together like the yarns and strands of a tow-line. Quick! I'll blow the fire."

When at last the twelve rods were made, Ahab tried them, one by one, by spiralling them, with his own hand, round a long, heavy iron bolt. "A flaw! Work that over again, Perth."

This done, Perth was about to begin welding the twelve into one, when Ahab stayed his hand, and said he would weld his own iron. As, then, with regular, gasping hems, he hammered on the anvil, Perth passing to him the glowing rods, one after the other, and the hard pressed forge shooting up its intense straight flame, the Parsee passed silently, and bowing over his head towards the fire, seemed invoking some curse or some blessing on the toil.

At last the shank, in one complete rod, received its final heat; and as Perth, to temper it, plunged it all hissing into the cask of water near by, the scalding steam shot up into Ahab's bent face.

"Would'st thou brand me, Perth?" wincing for a moment with the pain; "have I been but forging my own branding-iron, then?"

"Pray God, not that; yet I fear something, Captain Ahab. Is not this harpoon for the White Whale?"

"For the white fiend! But now for the barbs; thou must make them thyself, man. Here are my razors—the best of steel; here, and make the barbs sharp as the needle-sleet of the Icy Sea."

Fashioned at last into an arrowy shape, and welded by Perth to the shank, the steel soon pointed the end of the iron; and as the blacksmith was about giving the barbs their final heat, prior to tempering them, he cried to Ahab to place the water-cask near.

"No, no—no water for that; I want it of the true death-temper. Ahoy, there! Tashtego, Queequeg, Daggoo! What say ye, pagans! Will ye give me as much blood as will cover this barb?" holding it high up. A cluster of dark nods replied, Yes. Three punctures were made in the heathen flesh, and the White Whale's barbs were then tempered.

Now, mustering the spare poles from below, and selecting one of hickory, with the bark still investing it, Ahab fitted the end to the socket of the iron. A coil of new tow-line was then unwound, and some fathoms of it taken to the windlass, and stretched to a great tension. Pressing his foot upon it, till the rope hummed like a harp-string, then eagerly bending over it, and seeing no strandings, Ahab exclaimed, "Good! and now for the seizings."

At one extremity the rope was unstranded, and the separate spread yarns were all braided and woven round the socket of the harpoon; the pole was then driven hard up into the socket; from the lower end the rope was traced half way along the pole's length, and firmly secured so, with intertwistings of twine. This done, pole, iron, and rope—like the Three Fates—remained inseparable, and Ahab moodily stalked away with the weapon.

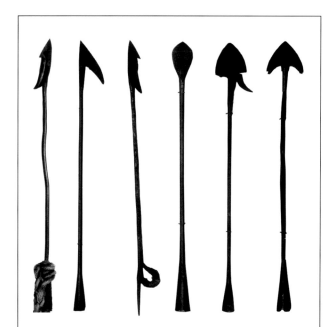

Weapons of the trade. From left to right: 1) Toggle harpoon, circa 1890, inscribed "BK CWM" (bark Charles W. Morgan); 2) one flued harpoon; 3) Temple toggle iron made with a darting-gun shaft (circa 1865-95) so it could be fired by Ebenezer Pierce's patented darting gun (1865); 4) lance harpoon; 5) explosive harpoon; 6) English harpoon.

Kendall Collection, NBWM

Chapter CXIV

The Gilder

In calm seas, boatsteerer Frank Rose guides the larboard boat from the *John R. Manta*, 1925. — *William R. Hegarty Collection.*

Penetrating further and further into the heart of the Japanese cruising ground, the Pequod was soon all astir in the fishery. Often, in mild, pleasant weather, for twelve, fifteen, eighteen, and twenty hours on the stretch, they were engaged in the boats, steadily pulling, or sailing, or paddling after the whales, or for an interlude of sixty or seventy minutes calmly awaiting their uprising; though with but small success for their pains.

At such times, under an abated sun; afloat all day upon smooth, slow heaving swells; seated in his boat, light as a birch canoe; and so sociably mixing with the soft waves themselves, that like hearth-stone cats they purr against the gunwale; these are the times of dreamy quietude, when beholding the tranquil beauty and brilliancy of the ocean's skin, one forgets the tiger heart that pants beneath it; and would not willingly remember, that this velvet paw but conceals a remorseless fang.

These are the times, when in his whale-boat the rover softly feels a certain filial, confident, land-like feeling towards the sea; that he regards it as so much flowery earth; and the distant ship revealing only the tops of her masts, seems struggling forward, not through high rolling waves, but through the tall grass of a rolling prairie: as when the western emigrants' horses only show their erected ears, while their hidden bodies widely wade through the amazing verdure.

The long-drawn virgin vales; the mild blue hill-sides; as over these there steals the hush, the hum; you almost swear that play-wearied children lie sleeping in these solitudes, in some glad May-time, when the flowers of the woods are plucked. And all this mixes with your most mystic mood; so that fact and fancy, half-way meeting, interpenetrate, and form one seamless whole.

Nor did such soothing scenes, however temporary, fail of at least as temporary an effect on Ahab. But if these secret golden keys did seem to open in him his own secret golden treasuries, yet did his breath upon them prove but tarnishing.

Calm before the storm, circa 1906. *The young, greenhand crew of the* John R. Manta *are on a calm-weather drill conducted by their captain, Henry S. Mandley—thus an unusually tranquil scene.* — *Photograph by Captain Henry S. Mandley.*

New Bedford Whaling Museum

Chapter cxv

The Pequod meets the Bachelor

Jolly enough were the sights and the sounds that came bearing down before the wind, some few weeks after Ahab's harpoon had been welded.

It was a Nantucket ship, the Bachelor, which had just wedged in her last cask of oil, and bolted down her bursting hatches; and now, in glad holiday apparel, was joyously, though somewhat vain-gloriously, sailing round among the widely-separated ships on the ground, previous to pointing her prow for home.

As was afterwards learned, the Bachelor had met with the most surprising success; all the more wonderful, for that while cruising in the same seas numerous other vessels had gone entire months without securing a single fish. Not only had barrels of beef and bread been given away to make room for the far more valuable sperm, but additional supplemental casks had been bartered for, from the ships she had met; and these were stowed along the deck, and in the captain's and officers' staterooms. Even the cabin table itself had been knocked into kindling-wood; and the cabin mess dined off the broad head of an oil-butt, lashed down to the floor for a centrepiece. In the forecastle, the sailors had actually caulked and pitched their chests, and filled them; it was humorously added, that the cook had clapped a head on his largest boiler, and filled it; that the steward had plugged his spare coffee-pot and filled it; that the harpooneers had headed the sockets of their irons and filled them; that indeed everything was filled with sperm, except the captain's pantaloons pockets, and those he reserved to thrust his hands into, in self-complacent testimony of his entire satisfaction.

As this glad ship of good luck bore down upon the moody Pequod, the barbarian sound of enormous drums came from her forecastle; and drawing still nearer, a crowd of her men were seen standing round her huge try-pots, which, covered with the parchment-like *poke* or stomach skin of the black fish, gave forth a loud roar to every stroke of the clenched hands of the crew. On the quarter-deck, the mates and harpooneers were dancing with the olive-hued girls who had eloped with them from the Polynesian Isles; while suspended in an ornamented boat, firmly secured aloft between the foremast and mainmast, three Long Island negroes, with glittering fiddle-bows of whale ivory, were presiding over the hilarious jig.

Lord and master over all this scene, the captain stood erect on the ship's elevated quarter-deck, so that the whole rejoicing drama was full before him,

Bark *Chili* meets bark *Malta* of New Bedford. "She [*Malta*] said they would take our oil for $1.00 per bbl, got it ready & took it on board of her. Also we sent several letters for home. I sent 2." — *From the journal of Rodolphus W. Dexter of Tisbury aboard bark* Chili *of New Bedford. Kendall Collection, NBWM.*

and seemed merely contrived for his own individual diversion. And Ahab, he too was standing on his quarter-deck, shaggy and black, with a stubborn gloom; and as the two ships crossed each other's wakes—one all jubilations for things passed, the other all forebodings as to things to come—their two captains in themselves impersonated the whole striking contrast of the scene.

"Come aboard, come aboard!" cried the gay Bachelor's commander, lifting a glass and a bottle in the air.

"Hast seen the White Whale?" gritted Ahab in reply.

"No; only heard of him; but don't believe in him at all," said the other good-humoredly. "Come aboard!"

"Thou are too damned jolly. Sail on. Hast lost any men?"

"Not enough to speak of—two islanders, that's all;—but come aboard, old hearty, come along. I'll soon take that black from your brow. Come along, will ye (merry's the play); a full ship and homeward-bound."

"How wondrous familiar is a fool!" muttered Ahab; then aloud, "Thou art a full ship and homeward bound, thou sayest; well, then, call me an empty ship, and outward-bound. So go thy ways, and I will mine. Forward there! Set all sail, and keep her to the wind!"

And thus, while the one ship went cheerily before the breeze, the other stubbornly fought against it; and so the two vessels parted; the crew of the Pequod looking with grave, lingering glances towards the receding Bachelor; but the Bachelor's men never heeding their gaze for the lively revelry they were in. And as Ahab, leaning over the taffrail, eyed the homeward-bound craft, he took from his pocket a small vial of sand, and then looking from the ship to the vial, seemed thereby bringing two remote associations together, for that vial was filled with Nantucket soundings.

Chapter CXVI

The Dying Whale

Not seldom in this life, when, on the right side, fortune's favorites sail close by us, we, though all adroop before, catch somewhat of the rushing breeze, and joyfully feel our bagging sails fill out. So seemed it with the Pequod. For next day after encountering the gay Bachelor, whales were seen and four were slain; and one of them by Ahab.

It was far down the afternoon; and when all the spearings of the crimson fight were done: and floating in the lovely sunset sea and sky, sun and whale both stilly died together; then, such a sweetness and such plaintiveness, such inwreathing orisons curled up in that rosy air, that it almost seemed as if far over from the deep green convent valleys of the Manilla isles, the Spanish land-breeze, wantonly turned sailor, had gone to sea, freighted with these vesper hymns.

Soothed again, but only soothed to deeper gloom, Ahab, who had sterned off from the whale, sat intently watching his final wanings from the now tranquil boat. For that strange spectacle observable in all sperm whales dying—the turning sunwards of the head, and so expiring—that strange spectacle, beheld of such a placid evening, somehow to Ahab conveyed a wondrousness unknown before.

Dying whales, 1846. — *Watercolor by George A. Gould aboard Columbia of Nantucket.*

"He turns and turns him to it,—how slowly, but how steadfastly, his homage-rendering and invoking brow, with his last dying motions. He too worships fire. Look! here, far water-locked; where for long Chinese ages, the billows have still rolled on speechless and unspoken to, as stars that shine upon the Niger's unknown source; here, too, life dies sunwards full of faith; but see! no sooner dead, than death whirls round the corpse, and it heads some other way.—

"Oh, thou dark Hindoo half of nature; thou art an infidel, thou queen, and too truly speakest to me in the wide-slaughtering Typhoon, and the hushed burial of its after calm. Yet dost thou, darker half, rock me with a prouder, if a darker faith. All thy unnamable imminglings, float beneath me here; I am buoyed by breaths of once living things, exhaled as air, but water now.

"Then hail, for ever hail, O sea, in whose eternal tossings the wild fowl finds his only rest. Born of earth, yet suckled by the sea; though hill and valley mothered me, ye billows are my foster-brothers!"

Dying whales before the Acushnet, circa 1846. *One of few authentic renderings of Melville's ship, this journal painting of Acushnet's crew killing a whale was made by boatsteerer Henry Johnson on the voyage subsequent to Melville's. "In vain, oh whale, dost thou seek intercedings with yon all-quickening sun, that only calls forth life, but gives it not again. Yet dost thou, darker half, rock me with a prouder, if a darker faith. All thy unnamable imminglings, float beneath me here; I am buoyed by breaths of once living things, exhaled as air, but water now."* — Ahab: Chapter 116.

Peabody Salem Museum

Chapter CXVII

The Whale Watch

"Boat Laying by a Dead Whale." — *Watercolor by John Bertoncini aboard schooner* Bonanza *of San Francisco, 1904. Kendall Collection, NBWM.*

The four whales slain that evening had died wide apart; one, far to windward; one, less distant, to leeward; one ahead; one astern. These last three were brought alongside ere nightfall; but the windward one could not be reached till morning; and the boat that had killed it lay by its side all night; and that boat was Ahab's.

The waif-pole was thrust upright into the dead whale's spout-hole; and the lantern hanging from its top, cast a troubled flickering glare upon the black, glossy back, and far out upon the midnight waves, which gently chafed the whale's broad flank, like soft surf upon a beach.

Ahab and all his boat's crew seemed asleep but the Parsee; who crouching in the bow, sat watching the sharks, that spectrally played round the whale, and tapped the light cedar planks with their tails. A sound like the moaning in squadrons over Asphaltites of unforgiven ghosts of Gomorrah, ran shuddering through the air.

Started from his slumbers, Ahab, face to face, saw the Parsee; and hooped round by the gloom of the night they seemed the last men in a flooded world. "I have dreamed it again," said he.

"Of the hearses? Have I not said, old man, that neither hearse nor coffin can be thine?"

"And who are hearsed that die on the sea?"

"But I said, old man, that ere thou couldst die on this voyage, two hearses must verily be seen by thee on the sea; the first not made by mortal hands; and the visible wood of the last one must be grown in America."

"Aye, aye! a strange sight that, Parsee:—a hearse and its plumes floating over the ocean with the waves for the pall-bearers. Ha! Such a sight we shall not soon see."

"Believe it or not, thou canst not die till it be seen, old man."

"And what was that saying about thyself?"

"Though it come to the last, I shall still go before thee thy pilot."

"And when thou art so gone before—if that ever befall—then ere I can follow, thou must still appear to me, to pilot me still?—Was it not so? Well, then, did I believe all ye say, oh my pilot! I have here two pledges that I shall yet slay Moby Dick and survive it."

"Take another pledge, old man," said the Parsee, as his eyes lighted up like fire-flies in the gloom—"Hemp only can kill thee."

"The gallows, ye mean.—I am immortal then, on land and on sea," cried Ahab, with a laugh of derision;— "Immortal on land and on sea!"

Both were silent again, as one man. The grey dawn came on, and the slumbering crew arose from the boat's bottom, and ere noon the dead whale was brought to the ship.

Whaling in the Atlantic, 1925. *Crew of the schooner* John R. Manta *pull alongside their whale.*

William R. Hegarty Collection

Chapter CXVIII

The Quadrant

The season for the Line at length drew near; and every day when Ahab, coming from his cabin, cast his eyes aloft, the vigilant helmsman would ostentatiously handle his spokes, and the eager mariners quickly run to the braces, and would stand there with all their eyes centrally fixed on the nailed doubloon; impatient for the order to point the ship's prow for the equator. In good time the order came. It was hard upon high noon; and Ahab, seated in the bows of his high-hoisted boat, was about taking his wonted daily observation of the sun to determine his latitude.

Now, in that Japanese sea, the days in summer are as freshets of effulgences. That unblinkingly vivid Japanese sun seems the blazing focus of the glassy ocean's immeasureable burning-glass. The sky looks lacquered; clouds there are none; the horizon floats; and this nakedness of unrelieved radiance is as the insufferable splendors of God's throne. Well that Ahab's quadrant was furnished with colored glasses, through which to take sight of that solar fire. So, swinging his seated form to the roll of the ship, and with his astrological-looking instrument placed to his eye, he remained in that posture for some moments to catch the precise instant when the sun should gain its precise meridian. Meantime while his whole attention was absorbed, the Parsee was kneeling beneath him on the ship's deck, and with face thrown up like Ahab's, was eyeing the same sun with him; only the lids of his eyes half hooded their orbs, and his wild face was subdued to an earthly passionlessness. At length the desired observation was taken; and with his pencil upon his ivory leg, Ahab soon calculated what his latitude must be at that precise instant. Then falling into a moment's revery, he again looked up towards the sun and murmured to himself: "Thou sea-mark! thou high and mighty Pilot! thou tellest me truly where I *am*—but canst thou cast the least hint where I *shall* be? Or canst thou tell where some other thing besides me is this moment living? Where is Moby Dick? This instant thou must be eyeing him. These eyes of mine look into the very eye that is even now beholding him; aye, and into the eye that is even now equally beholding the objects on the unknown, thither side of thee, thou sun!"

Then gazing at his quadrant, and handling, one after the other, its numerous cabalistical contrivances, he pondered again, and muttered: "Foolish toy! Science! Curse thee, thou vain toy; and cursed be all the things that cast

Shooting the sun from deck of the *John R. Manta*, 1925. — *William R. Hegarty Collection.*

man's eyes aloft to that heaven, whose live vividness but scorches him, as these old eyes are even now scorched with thy light, O sun! Level by nature to this earth's horizon are the glances of man's eyes; not shot from the crown of his head, as if God had meant him to gaze on his firmament. Curse thee, thou quadrant!" dashing it to the deck, "no longer will I guide my earthly way by thee; the level ship's compass, and the level dead-reckoning, by log and by line; *these* shall conduct me, and show me my place on the sea. Aye," lighting from the boat to the deck, "thus I trample on thee, thou paltry thing that feebly pointest on high; thus I split and destroy thee!"

As the frantic old man thus spoke and thus trampled with his live and dead feet, a sneering triumph that seemed meant for Ahab, and a fatalistic despair that seemed meant for himself—these passed over the mute, motionless Parsee's face. Unobserved he rose and glided away; while, awestruck by the aspect of their commander, the seamen clustered together on the forecastle, till Ahab, troubledly pacing the deck, shouted out—"To the braces! Up helm!—square in!"

In an instant the yards swung round; and as the ship half-wheeled upon her heel, her three firm-seated graceful masts erectly poised upon her long, ribbed hull, seemed as the three Horatii pirouetting on one sufficient steed.

Standing between the knight-heads, Starbuck watched the Pequod's tumultuous way, and Ahab's also, as he went lurching along the deck.

"I have sat before the dense coal fire and watched it all aglow, full of its tormented flaming life; and I have seen it wane at last, down, down, to dumbest dust. Old man of oceans! of all this fiery life of thine, what will at length remain but one little heap of ashes!"

"Aye," cried Stubb, "but sea-coal ashes—mind ye that, Mr. Starbuck—sea-coal, not your common charcoal. Well, well; I heard Ahab mutter, 'Here some one thrusts these cards into these old hands of mine; swears that I must play them, and no others.' And damn me, Ahab, but thou actest right; live in the game, and die it!"

Chapter CXIX

The Candles

Warmest climes but nurse the cruellest fangs: the tiger of Bengal crouches in spiced groves of ceaseless verdure. So, too, it is, that in these resplendent Japanese seas the mariner encounters the direst of all storms, the Typhoon. It will sometimes burst from out that cloudless sky, like an exploding bomb upon a dazed and sleepy town.

The British whaling ship *William* of London weathers a stiff gale in the South Pacific, 1796. — *Watercolor by crewman Thomas Wetling.*

Towards evening of that day, the Pequod was torn of her canvas, and bare-poled was left to fight a Typhoon which had struck her directly ahead. When darkness came on, sky and sea roared and split with the thunder, and blazed with the lightning, that showed the disabled masts fluttering here and there with the rags which the first fury of the tempest had left for its after sport.

Holding by a shroud, Starbuck was standing on the quarter-deck; at every flash of the lightning glancing aloft, to see what additional disaster might have befallen the intricate hamper there; while Stubb and Flask were directing the men in the higher hoisting and firmer lashing of the boats. But all their pains seemed naught. Though lifted to the very top of the cranes, the windward quarter boat (Ahab's) did not escape. A great rolling sea, dashing high up against the reeling ship's high tetering side, stove in the boat's bottom at the stern, and left it again, all dripping through like a sieve.

"Bad work, bad work! Mr. Starbuck," said Stubb, regarding the wreck, "but the sea will have its way. But never mind; it's all in fun: so the old song says;"—(*sings*).

"Avast Stubb," cried Starbuck, "let the Typhoon sing, and strike his harp here in our rigging; but if thou art a brave man thou wilt hold thy peace."

"But I am not a brave man; never said I was a brave man; I am a coward; and I sing to keep up my spirits. And I tell you what it is, Mr. Starbuck, there's no way to stop my singing in this world but to cut my throat. And when that's done, ten to one I sing ye the doxology for a wind-up."

"Madman! look through my eyes if thou hast none of thine own."

"What! how can you see better of a dark night than anybody else, never mind how foolish?"

"Here!" cried Starbuck, seizing Stubb by the shoulder, and pointing his hand towards the weather bow, "markest thou not that the gale comes from the eastward, the very course Ahab is to run for Moby Dick? the very course he swung to this day noon? now mark his boat there; where is that stove? In the stern-sheets, man; where he is wont to stand—his stand-point is stove, man! Now jump overboard, and sing away, if thou must!"

"I don't half understand ye: what's in the wind?"

"Yes, yes, round the Cape of Good Hope is the shortest way to Nantucket," soliloquized Starbuck suddenly, heedless of Stubb's question. "The gale that now hammers at us to stave us, we can turn it into a fair wind that will drive us towards home. Yonder, to windward, all is blackness of doom; but to leeward, homeward—I see it lightens up there; but not with the lightning."

Now, as the lightning rod to a spire on shore is intended to carry off the perilous fluid into the soil; so the kindred rod which at sea some ships carry to each mast, is intended to conduct it into the water. But as this conductor must descend to considerable depth, that its end may avoid all contact with the hull; and as moreover, if kept constantly towing there, it would be liable to many mishaps, besides interfering not a little with some of the rigging, and more or less impeding the vessel's way in the water; because of all this, the lower parts of a ship's lightning-rods are not always overboard; but are generally made in long slender links, so as to be the more readily hauled up into the chains outside, or thrown down into the sea, as occasion may require.

"The rods! the rods!" cried Starbuck to the crew, suddenly admonished to vigilance by the vivid lightning that had just been darting flambeaux, to light Ahab to his post. "Are they overboard? drop them over, fore and aft. Quick!

"Look aloft!" cried Starbuck. "Corpusants! the corpusants!"

All the yard-arms were tipped with a pallid fire; and touched at each tri-pointed lightning-rod-end with three tapering white flames, each of the three tall masts was silently burning in that sulphurous air, like three gigantic wax tapers before an altar.

While this pallidness was burning aloft, few words were heard from the enchanted crew; who in one thick cluster stood on the forecastle, all their eyes gleaming in that pale phosphorescence, like a far away constellation of stars.

"Aye, aye, men!" cried Ahab. "Look up at it; mark it well; the white flame but lights the way to the White Whale! Hand me those main-mast links there; I would fain feel this pulse, and let mine beat against it; blood against fire! So." Then turning—the last link held fast in his left hand, he put his foot upon the Parsee; and with fixed upward eye, and high-flung right arm, he stood erect before the lofty tri-pointed trinity of flames.

"The boat! the boat!" cried Starbuck, "look at thy boat, old man!"

Ahab's harpoon, the one forged at Perth's fire, remained firmly lashed in its conspicuous crotch, so that it projected beyond his whale-boat's bow; but the sea that had stove its bottom had caused the loose leather sheath to drop off; and from the keen steel barb there now came a levelled flame of pale, forked fire. As the silent harpoon burned there like a serpent's tongue, Starbuck grasped Ahab by the arm—"God, God is against thee, old man; forbear! 'Tis an ill voyage! ill begun, ill continued; let me square the yards, while we may, old man, and make a fair wind of it homewards, to go on a better voyage than this."

Overhearing Starbuck, the panic-stricken crew instantly ran to the braces—though not a sail was left aloft. For the moment all the aghast mate's thoughts seemed theirs; they raised a half mutinous cry. But dashing the rattling lightning links to the deck, and snatching the burning harpoon, Ahab waved it like a torch among them; swearing to transfix with it the first sailor that but cast loose a rope's end. Petrified by his aspect, and still more shrinking from the fiery dart that he held, the men fell back in dismay, and Ahab again spoke:—

"All your oaths to hunt the White Whale are as binding as mine; and heart, soul, and body, lungs and life, old Ahab is bound. And that ye may know to what tune this heart beats; look ye here; thus I blow out the last fear!" And with one blast of his breath he extinguished the flame.

As in the hurricane that sweeps the plain, men fly the neighborhood of some lone, gigantic elm, whose very height and strength but render it so much the more unsafe, because so much the more a mark for thunderbolts; so at those last words of Ahab's many of the mariners did run from him in a terror of dismay.

Chelsea *sailing on her beam-ends in lightning gale, 1873.* *"Hear me, then: I take that mast-head flame we saw for a sign of good luck; for those masts are rooted in a hold that is going to be chock a' block with sperm-oil, d'ye see; and so, all that sperm will work up into the masts, like sap in a tree. Yes, our three masts will yet be as three spermaceti candles—that's the good promise we saw."* — *Stubb, Chapter 119. Engraving from Davis:* Nimrod of the Sea.

Chapter CXXIII

The Musket

During the most violent shocks of the Typhoon, the man at the Pequod's jaw-bone tiller had several times been reelingly hurled to the deck by its spasmodic motions, even though preventer tackles had been attached to it—for they were slack—because some play to the tiller was indispensable.

In a severe gale like this, while the ship is but a tossed shuttle-cock to the blast, it is by no means uncommon to see the needles in the compasses, at intervals, go round and round. It was thus with the Pequod's; at almost every shock the helmsman had not failed to notice the whirling velocity with which they revolved upon the cards; it is a sight that hardly any one can behold without some sort of unwonted emotion. Some hours after midnight, the Typhoon abated so much, that through the strenuous exertions of Starbuck and Stubb—one engaged forward and the other aft—the shivered remnants of the jib and fore and main-top-sails were cut adrift from the spars, and went eddying away to leeward, like the feathers of an albatross, which sometimes are cast to the winds when that storm-tossed bird is on the wing.

The three corresponding new sails were now bent and reefed, and a storm-trysail was set further aft; so that the ship soon went through the water with some precision again; and the course—for the present, East-south-east—which he was to steer, if practicable, was once more given to the helmsman. For during the violence of the gale, he had only steered according to its vicissitudes. But as he was now bringing the ship as near her course as possible, watching the compass meanwhile, lo! a good sign! the wind seemed coming round astern; aye! the foul breeze became fair!

Instantly the yards were squared, to the lively song of *"Ho! the fair wind! oh-he-yo, cheerly, men!"* the crew singing for joy, that so promising an event should so soon have falsified the evil portents preceding it.

In compliance with the standing order of his commander—to report immediately, and at any one of the twenty-four hours, any decided change in the affairs of the deck,—Starbuck had no sooner trimmed the yards to the breeze—however reluctantly and gloomily,—than he mechanically went below to apprise Captain Ahab of the circumstance.

Ere knocking at his state-room, he involuntarily paused before it a moment. The cabin lamp—taking long swings this way and that—was burning fitfully, and casting fitful shadows upon the old man's bolted door. The isolated subterraneousness of the cabin made a certain humming silence to reign there, though it was hooped round by all the roar of the elements. The loaded muskets in the rack were shiningly revealed, as they stood upright against the forward bulkhead. Starbuck was an honest, upright man; but out of Starbuck's heart, at that instant when he saw the muskets, there strangely evolved an evil thought; but so blent with its neutral or good accompaniments that for the instant he hardly knew it for itself.

"He would have shot me once," he murmured, "yes, there's the very musket that he pointed at me;—that one with the studded stock; let me touch it—lift it. Strange, that I, who have handled so many deadly lances, strange, that I should shake so now. Loaded? I must see. Aye, aye; and powder in the pan;—that's not good. Best spill it?—wait. I'll cure myself of this. I'll hold the musket boldly while I think.—I come to report a fair wind to him. But how fair? Fair for death and doom,—*that's* fair for Moby Dick. It's a fair wind that's only fair for that accursed fish.—The very tube he pointed at me!—the very one; *this* one—I hold it here; he would have killed me with the very thing I handle now.—Aye and he would fain kill all his crew. Does he not say he will not strike his spars to any gale? Has he not dashed his heavenly quadrant? But shall this crazed old man be tamely suffered to drag a whole ship's company down to doom with him?—Yes, it would make him the wilful murderer of thirty men and more, if this ship come to any deadly harm; and come to deadly harm, my soul swears this ship will, if Ahab have his way. If, then, he were this instant—put aside, that crime would not be his. Ha! is he muttering in his sleep? Yes, just there,—in there, he's sleeping.

Nantucket whaleship *Globe* marooned at Mulgrave Islands following her crew's horrific mutiny in 1824.
— *From Comstock:* The Life of Samuel Comstock…, *1840.*

Sleeping? aye, but still alive, and soon awake again. I can't withstand thee, then, old man. Not reasoning; not remonstrance; not entreaty wilt thou hearken to; all this thou scornest. Flat obedience to thy own flat commands, this is all thou breathest. Aye, and say'st the men have vow'd thy vow; say'st all of us are Ahabs. Great God forbid!—But is there no other way? no lawful way?—Make him a prisoner to be taken home? What! hope to wrest this old man's living power from his own living hands? Only a fool would try it. Say he were pinioned even; knotted all over with ropes and hawsers; chained down to ring-bolts on this cabin floor; he would be more hideous than a caged tiger, then. I could not endure the sight; could not possibly fly his howlings; all comfort, sleep itself, inestimable reason would leave me on the long intolerable voyage. What, then, remains? The land is hundreds of leagues away, and locked Japan the nearest. I stand alone here upon an open sea, with two oceans and a whole continent between me and law.—Aye, aye, 'tis so.—Is heaven a murderer when its lightning strikes a would-be murderer in his bed, tindering sheets and skin together?—And would I be a murderer, then, if—" and slowly, stealthily, and half sideways looking, he placed the loaded musket's end against the door.

"On this level, Ahab's hammock swings within; his head this way. A touch, and Starbuck may survive to hug his wife and child again.—Oh Mary! Mary!—boy! boy! boy!—But if I wake thee not to death, old man, who can tell to what unsounded deeps Starbuck's body this day week may sink, with all the crew! Great God, where art thou? Shall I? shall I?—The wind has gone down and shifted, sir; the fore and main topsails are reefed and set; she heads her course."

"Stern all! Oh Moby Dick, I clutch thy heart at last!"

Such were the sounds that now came hurtling from out the old man's tormented sleep, as if Starbuck's voice had caused the long dumb dream to speak.

The yet levelled musket shook like a drunkard's arm against the panel; Starbuck seemed wrestling with an angel; but turning from the door, he placed the death-tube in its rack, and left the place.

"He's too sound asleep, Mr. Stubb; go thou down, and wake him, and tell him. I must see to the deck here. Thou know'st what to say."

Mutiny on the Sharon, circa 1845. *Starbuck's mutinous feelings toward Ahab could have been taken from Melville's own experiences. In September 1842, near Tahiti, Melville and six shipmates were charged with mutiny and shackled in irons for refusing duty on the Lucy Ann. Just three months later, the Polynesian crew of the Fairhaven ship Sharon staged a mutiny and killed their abusive captain, Howes Norris. The Sharon mutiny, like the Globe mutiny, was a well-known story of the day. The scene depicted here shows the rest of the crew returning from a whale hunt to find the Pacific Islanders in command of the ship. The American crew eventually retook control of the vessel.* — Sources: Ellis, Kugler. From Russell and Purrington's panorama.

New Bedford Whaling Museum

Chapter CXXIV

The Needle

Next morning the not-yet-subsided sea rolled in long slow billows of mighty bulk, and striving in the Pequod's gurgling track, pushed her on like giants' palms outspread. The strong, unstaggering breeze abounded so, that sky and air seemed vast outbellying sails; the whole world boomed before the wind. The sea was as a crucible of molten gold, that bubblingly leaps with light and heat.

Long maintaining an enchanted silence, Ahab stood apart; and every time the teetering ship loweringly pitched down her bowsprit, he turned to eye the bright sun's rays produced ahead; and when she profoundly settled by the stern, he turned behind, and saw the sun's rearward place, and how the same yellow rays were blending with his undeviating wake.

But suddenly reined back by some counter thought, he hurried towards the helm, huskily demanding how the ship was heading.

"East—sou-east, Sir," said the frightened steersman.

"Thou liest!" smiting him with his clenched fist. "Heading East at this hour in the morning, and the sun astern?"

Standing behind him Starbuck looked, and lo! the two compasses pointed East, and the Pequod was as infallibly going West.

But ere the first wild alarm could get out abroad among the crew, the old man with a rigid laugh exclaimed, "I have it! It has happened before. Mr. Starbuck, last night's thunder turned our compasses—that's all."

Here, it must needs be said, that accidents like this have in more than one case occurred to ships in violent storms. The magnetic energy, as developed in the mariner's needle, is, as all know, essentially one with the electricity beheld in heaven; hence it is not to be much marvelled at, that such things should be. In instances where the lightning has actually struck the vessel, so as to smite down some of the spars and rigging, the effect upon the needle has at times been still more fatal; all its loadstone virtue being annihilated, so that the before magnetic steel was of no more use than an old wife's knitting needle. But in either case, the needle never again, of itself, recovers the original virtue thus marred or lost; and if the binnacle compasses be affected, the same fate reaches all the others that may be in the ship; even were the lowermost one inserted into the kelson.

Deliberately standing before the binnacle, and eyeing the transpointed compasses, the old man, with the sharp of his extended hand, now took the precise bearing of the sun, and satisfied that the needles were exactly inverted, shouted out his orders for the ship's course to be changed accordingly.

Accessory, perhaps, to the impulse dictating the thing he was now about to do, were certain prudential motives, whose object might have been to revive the spirits of his crew by a stroke of his subtile skill, in a matter so wondrous as that of the inverted compasses. Besides, the old man well knew that to steer by transpointed needles, though clumsily practicable, was not a thing to be passed over by superstitious sailors, without some shudderings and evil portents.

"Men," said he, steadily turning upon the crew, as the mate handed him the things he had demanded, "my men, the thunder turned old Ahab's needles; but out of this bit of steel Ahab can make one of his own, that will point as true as any."

Abashed glances of servile wonder were exchanged by the sailors, as this was said; and with fascinated eyes they awaited whatever magic might follow. But Starbuck looked away.

With a blow from the top-maul Ahab knocked off the steel head of the lance, and then handing to the mate the long iron rod remaining, bade him hold it upright, without its touching the deck. Then, with the maul, after repeatedly smiting the upper end of this iron rod, he placed the blunted needle endwise on the top of it, and less strongly hammered that, several times, the mate still holding the rod as before. Then going through some small strange motions with it—whether indispensable to the magnetizing of the steel, or merely intended to augment the awe of the crew, is uncertain—he called for linen thread; and moving to the binnacle, slipped out the two reversed needles there, and horizontally suspended the sail-needle by its middle, over one of the compass-cards. At first, the steel went round and round, quivering and vibrating at either end; but at last it settled to its place, when Ahab, who had been intently watching for this result, stepped frankly back from the binnacle, and pointing his stretched arm towards it, exclaimed,—"Look ye, for yourselves, if Ahab be not the lord of the level loadstone! The sun is East, and that compass swears it!"

One after another they peered in, for nothing but their own eyes could persuade such ignorance as theirs, and one after another they slunk away.

In his fiery eyes of scorn and triumph, you then saw Ahab in all his fatal pride.

Surveyor's compass made by James Halsey, Boston, circa 1850.
— *Kendall Collection, NBWM.*

Chapter CXXVI

The Life Buoy

Hood's Island, New Zealand, 1843: site of a gam between the ship *Columbia* of Nantucket and the *Eagle* of New Bedford. — *From the journal of George Gould aboard the* Columbia. *Kendall Collection, NBWM.*

Steering now south-eastward by Ahab's levelled steel, and her progress solely determined by Ahab's level log and line; the Pequod held on her path towards the Equator. Making so long a passage through such unfrequented waters, descrying no ships, and ere long, sideways impelled by unvarying trade winds, over waves monotonously mild; all these seemed the strange calm things preluding some riotous and desperate scene.

At last, when the ship drew near to the outskirts, as it were, of the Equatorial fishing-ground, and in the deep darkness that goes before the dawn, was sailing by a cluster of rocky islets; the watch was startled by a cry so plaintively wild and unearthly that one and all, they started from their reveries, and for the space of some moments stood, or sat, or leaned all transfixedly listening, like the carved Roman slave, while that wild cry remained within hearing. The Christian or civilized part of the crew said it was mermaids, and shuddered; but the pagan harpooneers remained unappalled. Yet the grey Manxman—the oldest mariner of all—declared that the wild thrilling sounds that were heard, were the voices of newly drowned men in the sea.

Below in his hammock, Ahab did not hear of this till grey dawn, when he came to the deck; it was then recounted to him by Flask, not unaccompanied with hinted dark meanings. He hollowly laughed, and thus explained the wonder.

Those rocky islands the ship had passed were the resort of great numbers of seals, and some young seals that had lost their dams, or some dams that had lost their cubs, must have risen nigh the ship and kept company with her, crying and sobbing with their human sort

A whaleship passes a school of dolphins near the island of Dominique in the Marquesas, 1845. "My mind often reverts to the many pleasant moonlight watches we passed together on the deck of the *Acushnet* as we whiled away the hours with yarn and song till 'eight bells.'" — *Richard Tobias Greene in a letter to Melville, Apr 8, 1861 (Source: Leyda). Painting from Russell and Purrington's panorama.*

New Bedford Whaling Museum

A group of seals watches whaleships rounding Cape Horn, 1845.
— From Russell and Purrington's panorama. New Bedford Whaling Museum.

of wail. But this only the more affected some of them, because most mariners cherish a very superstitious feeling about seals, arising not only from their peculiar tones when in distress, but also from the human look of their round heads and semi-intelligent faces, seen peeringly uprising from the water alongside. In the sea, under certain circumstances, seals have more than once been mistaken for men.

But the bodings of the crew were destined to receive a most plausible confirmation in the fate of one of their number that morning. At sun-rise this man went from his hammock to his mast-head at the fore; and whether it was that he was not yet half waked from his sleep (for sailors sometimes go aloft in a transition state), whether it was thus with the man, there is now no telling; but, be that as it may, he had not been long at his perch, when a cry was heard—a cry and a rushing—and looking up, they saw a falling phantom in the air; and looking down, a little tossed heap of white bubbles in the blue of the sea.

The life-buoy—a long slender cask—was dropped from the stern, where it always hung obedient to a cunning spring; but no hand rose to seize it, and the sun having long beat upon this cask it had shrunken, so that it slowly filled, and the parched wood also filled at its every pore; and the studded iron-bound cask followed the sailor to the bottom, as if to yield him his pillow, though in sooth but a hard one.

And thus the first man of the Pequod that mounted the mast to look out for the White Whale, on the White Whale's own peculiar ground; that man was swallowed up in the deep.

The lost life-buoy was now to be replaced; Starbuck was directed to see to it; but as no cask of sufficient lightness could be found, and as in the feverish eagerness of what seemed the approaching crisis of the voyage, all hands were impatient of any toil but what was directly connected with its final end, whatever that might prove to be; therefore, they were going to leave the ship's stern unprovided with a buoy, when by certain strange signs and inuendoes Queequeg hinted a hint concerning his coffin.

"A life-buoy of a coffin!" cried Starbuck, starting.

"Rather queer, that, I should say," said Stubb.

"It will make a good enough one," said Flask, "the carpenter here can arrange it easily."

"Bring it up; there's nothing else for it," said Starbuck, after a melancholy pause. "Rig it, carpenter; do not look at me so—the coffin, I mean. Dost thou hear me? Rig it."

"And shall I nail down the lid, Sir?" moving his hand as with a hammer.

"Aye."

"And shall I caulk the seams, Sir?" moving his hand as with a caulking-iron.

"Aye."

"And shall I then pay over the same with pitch, Sir?" moving his hand as with a pitch-pot.

"Away! what possesses thee to this? Make a life-buoy of the coffin, and no more.—Mr. Stubb, Mr. Flask, come forward with me."

"Now I don't like this. I make a leg for Captain Ahab, and he wears it like a gentleman; but I make a bandbox for Queequeg, and he won't put his head into it. Are all my pains to go for nothing with that coffin? And now I'm ordered to make a life-buoy of it. It's like turning an old coat; going to bring the flesh on the other side now. I don't like this cobbling sort of business—I don't like it at all; it's undignified; it's not my place.

"Let me see. Nail down the lid; caulk the seams; pay over the same with pitch; batten them down tight, and hang it with the snap-spring over the ship's stern. Were ever such things done before with a coffin? Hem! I'll do the job, now, tenderly. I'll have me—let's see—how many in the ship's company, all told? But I've forgotten. Any way, I'll have me thirty separate, Turk's-headed life-lines, each three feet long hanging all round to the coffin. Then, if the hull go down, there'll be thirty lively fellows all fighting for one coffin, a sight not seen very often beneath the sun! Come hammer, calking-iron, pitch-pot, and marling-spike! Let's to it."

Queequeg's coffin, 1930. — Illustration by Rockwell Kent.

©1930 by R. R. Donnelley & Sons Co., with permission

Chapter CXXVIII

The Pequod Meets the Rachel

Next day, a large ship, the Rachel, was descried, bearing directly down upon the Pequod, all her spars thickly clustering with men. At the time the Pequod was making good speed through the water; but as the broad-winged windward stranger shot nigh to her, the boastful sails all fell together as blank bladders that are burst, and all life fled from the smitten hull.

"Bad news; she brings bad news," muttered the old Manxman. But ere her commander, who, with trumpet to mouth, stood up in his boat; ere he could hopefully hail, Ahab's voice was heard.

"Hast seen the White Whale?"

"Aye, yesterday. Have ye seen a whale-boat adrift?"

Throttling his joy, Ahab negatively answered this unexpected question; and would then have fain boarded the stranger, when the stranger captain himself, having

A large, unidentified whaleship, circa 1900.
— W. Tripp photograph. Kendall Collection, NBWM.

stopped his vessel's way, was seen descending her side. A few keen pulls, and his boat-hook soon clinched the Pequod's main-chains, and he sprang to the deck.

"Where was he?—not killed!—not killed!" cried Ahab, closely advancing. "How was it?"

It seemed that somewhat late on the afternoon of the day previous, while three of the stranger's boats were engaged with a shoal of whales, which had led them some four or five miles from the ship; and while they were yet in swift chase to windward, the white hump and head of Moby Dick had suddenly loomed up out of the blue water, not very far to leeward; whereupon, the fourth rigged boat—a reserved one—had been instantly lowered in chase. After a keen sail before the wind, this fourth boat—the swiftest keeled of all—seemed to have succeeded in fastening—at least, as well as the man at the mast-head could tell anything about it. In the distance he saw the diminished dotted boat; and then a swift gleam of bubbling white water; and after that nothing more; whence it was concluded that the stricken whale must have indefinitely run away with his pursuers, as often happens. There was some apprehension, but no positive alarm, as yet. The recall signals were placed in the rigging; darkness came on; and forced to pick up her three far to windward boats—ere going in quest of the fourth one in the precisely opposite direction—the ship had not only been necessitated to leave that boat to its fate till near midnight, but, for the time, to increase her distance from it. But the rest of her crew being at last safe aboard, she crowded all sail—stunsail on stunsail—after the missing boat; kindling a fire in her try-pots for a beacon; and every other man aloft on the look-out. But though when she had thus sailed a sufficient distance to gain the presumed place of the absent ones when last seen; though she then paused to lower her spare boats to pull all around her; and not finding anything, had again dashed on; again paused, and lowered her boats; and though she had thus continued doing till day light; yet not the least glimpse of the missing keel had been seen.

The story told, the stranger Captain immediately went on to reveal his object in boarding the Pequod. He desired that ship to unite with his own in the search.

Bark Desdemona, 1896. *Bound for Hudson Bay, Desdemona was lost off Whale Point, Nova Scotia just two months into this voyage. Her crew and 2,500 pounds of whalebone were saved.*

William R. Hegarty Collection

"My boy, my own boy is among them. For God's sake, I beg, I conjure—" here exclaimed the stranger Captain to Ahab, who thus far had but icily received his petition. "For eight-and-forty hours let me charter your ship—I will gladly pay for it, and roundly pay for it—if there be no other way—for eight-and-forty hours only—only that—you must, oh, you must, and you *shall* do this thing."

"His son!" cried Stubb, "oh, it's his son he's lost! I take back the coat and watch—what says Ahab? We must save that boy."

"He's drowned with the rest on 'em, last night," said the old Manx sailor standing behind them; "I heard; all of ye heard their spirits."

Meantime, the stranger was still beseeching his poor boon of Ahab; and Ahab still stood like an anvil, receiving every shock, but without the least quivering of his own.

"Avast," cried Ahab; then in a voice that prolongingly moulded every word—"Captain Gardiner, I will not do it. Even now I lose time. Good bye, good bye. God bless ye, man, and may I forgive myself, but I must go. Mr. Starbuck, look at the binnacle watch, and in three minutes from this present instant warn off all strangers: then brace forward again, and let the ship sail as before."

Hurriedly turning, with averted face, he descended into his cabin, leaving the strange captain transfixed at this unconditional and utter rejection of his so earnest suit. But starting from his enchantment, Gardiner silently hurried to the side; more fell than stepped into his boat, and returned to his ship.

Soon the two ships diverged their wakes; and long as the strange vessel was in view, she was seen to yaw hither and thither at every dark spot, however small, on the sea. This way and that her yards were swung round; starboard and larboard, she continued to tack; now she beat against a head sea; and again it pushed her before it; while all the while, her masts and yards were thickly clustered with men, as three tall cherry trees, when the boys are cherrying among the boughs.

But by her still halting course and winding, woeful way, you plainly saw that this ship that so wept with spray, still remained without comfort. She was Rachel, weeping for her children, because they were not.

Bark Plantina, circa 1910. *"I will wager something now," whispered Stubb to Flask, "that some one in that missing boat wore off that Captain's best coat; mayhap, his watch—he's so cursed anxious to get it back. Who ever heard of two pious whale-ships cruising after one missing whale-boat in the height of the whaling season?"* — Chapter 128. George Nye photograph.

Kendall Collection, NBWM

Chapter cxxxi

The Pequod Meets the Delight

The intense Pequod sailed on; the rolling waves and days went by; the life-buoy-coffin still lightly swung; and another ship, most miserably misnamed the Delight, was descried. As she drew nigh, all eyes were fixed upon her broad beams, called shears, which, in some whaling-ships, cross the quarter-deck at the height of eight or nine feet; serving to carry the spare, unrigged, or disabled boats.

Upon the stranger's shears were beheld the shattered, white ribs, and some few splintered planks, of what had once been a whale-boat; but you now saw

Whaleships passing near Rock of Gibraltor, circa 1843.
— From the journal of John Francis Akin aboard ship Virginia of New Bedford.

through this wreck, as plainly as you see through the peeled, half-unhinged, and bleaching skeleton of a horse.

"Hast seen the White Whale?"

"Look!" replied the hollow-cheeked captain from his taffrail; and with his trumpet he pointed to the wreck.

"Hast killed him?"

"The harpoon is not yet forged that will ever do that," answered the other, sadly glancing upon a rounded hammock on the deck, whose gathered sides some noiseless sailors were busy in sewing together.

"Not forged!" and snatching Perth's levelled iron from the crotch, Ahab held it out, exclaiming—"Look ye, Nantucketer; here in this hand I hold his death! Tempered in blood, and tempered by lightning are these barbs; and I swear to temper them triply in that hot place behind the fin, where the White Whale most feels his accursed life!"

"Then God keep thee, old man—see'st thou that"—pointing to the hammock—"I bury but one of five stout men, who were alive only yesterday; but were dead ere night. Only *that* one I bury; the rest were buried before they died; you sail upon their tomb." Then turning to his crew—"Are ye ready there? place the plank then on the rail, and lift the body; so, then—Oh! God"—advancing towards the hammock with uplifted hands—"may the resurrection and the life—"

"Brace forward! Up helm!" cried Ahab like lightning to his men.

But the suddenly started Pequod was not quick enough to escape the sound of the splash that the corpse soon made as it struck the sea; not so quick, indeed, but that some of the flying bubbles might have sprinkled her hull with their ghostly baptism.

As Ahab now glided from the dejected Delight, the strange life-buoy hanging at the Pequod's stern came into conspicuous relief.

"Ha! yonder! look yonder, men!" cried a foreboding voice in her wake. "In vain, oh, ye strangers, ye fly our sad burial; ye but turn us your taffrail to show us your coffin!"

Bark Canton, full sail, circa 1900. Built in 1835, Canton was the oldest whaleship in the New Bedford fleet at the time of this photograph. She was lost off Cape Verde Islands in 1910, but all hands were saved.

William R. Hegarty Collection

Chapter CXXXII

The Symphony

Ship *Constantine*, 1849. — *From journal of Owen Chase, Jr.*

t was a clear steel-blue day. The firmaments of air and sea were hardly separable in that all-pervading azure; only, the pensive air was transparently pure and soft, with a woman's look, and the robust and man-like sea heaved with long, strong, lingering swells, as Samson's chest in his sleep.

Hither, and thither, on high, glided the snow-white wings of small, unspeckled birds; these were the gentle thoughts of the feminine air; but to and fro in the deeps, far down in the bottomless blue, rushed mighty leviathans, sword-fish, and sharks; and these were the strong, troubled, murderous thinkings of the masculine sea.

Tied up and twisted; gnarled and knotted with wrinkles; haggardly firm and unyielding; his eyes glowing like coals, that still glow in the ashes of ruin; untottering Ahab stood forth in the clearness of the morn; lifting his splintered helmet of a brow to the fair girl's forehead of heaven.

Slowly crossing the deck from the scuttle, Ahab leaned over the side, and watched how his shadow in the water sank and sank to his gaze, the more and the more that he strove to pierce the profundity. But the lovely aromas in that enchanted air did at last seem to dispel, for a moment, the cankerous thing in his soul. From beneath his slouched hat Ahab dropped a tear into the sea; nor did all the Pacific contain such wealth as that one wee drop.

Starbuck saw the old man; saw him, how he heavily leaned over th e side; and he seemed to hear in his own true heart the measureless sobbing that stole out of the centre of the serenity around. Careful not to touch him, or be noticed by him, he yet drew near to him, and stood there.

Ahab turned.

"Starbuck!"

"Sir."

"Oh, Starbuck! it is a mild, mild wind, and a mild looking sky. On such a day—very much such a sweetness as this—I struck my first whale—a boy-harpooneer of

Bark A. R. Tucker, fully laden, returns home, 1906. *On her last voyage, the 145-ton* A. R. Tucker, *under Captain Sylvester B. Potter, brought home 1,670 barrels of sperm oil from the Atlantic.* — *Joseph G. Tirrell photograph.*

Kingman Family Collection

Bark Swallow, *circa 1900.* — *Joseph G. Tirrell photograph. Spinner Collection.*

eighteen! Forty—forty—forty years ago!—ago! Forty years of continual whaling! forty years of privation, and peril, and storm-time! forty years on the pitiless sea! for forty years has Ahab forsaken the peaceful land, for forty years to make war on the horrors of the deep! Aye and yes, Starbuck, out of those forty years I have not spent three ashore. Away, whole oceans away, from that young girl-wife I wedded past fifty, and sailed for Cape Horn the next day, leaving but one dent in my marriage pillow—wife? wife?—rather a widow with her husband alive! Aye, I widowed that poor girl when I married her, Starbuck;—aye, aye! what a forty years' fool—fool—old fool, has old Ahab been! Why this strife of the chase? why weary, and palsy the arm at the oar, and the iron, and the lance? how the richer or better is Ahab now? I feel deadly faint, bowed, and humped, as though I were Adam, staggering beneath the piled centuries since Paradise. Close! stand close to me, Starbuck; let me look into a human eye; it is better than to gaze into sea or sky; better than to gaze upon God."

"Oh, my Captain! my Captain! noble soul! grand old heart, after all! why should any one give chase to that hated fish! Away with me! let us fly these deadly waters! let us home! Wife and child, too, are Starbuck's—wife and child of his brotherly, sisterly, play-fellow youth; even as thine, sir, are the wife and child of thy loving, longing, paternal old age! Away! let us away!—this instant let me alter the course! Yes, yes! no more! it is done! we head for Nantucket! Come, my Captain, study out the course, and let us away! See, see! the boy's face from the window! the boy's hand on the hill!"

But Ahab's glance was averted; like a blighted fruit tree he shook, and cast his last, cindered apple to the soil.

"What is it, what nameless, inscrutable, unearthly thing is it; what cozzening, hidden lord and master, and cruel, remorseless emperor commands me; that against all natural lovings and longings, I so keep pushing, and crowding, and jamming myself on all the time; recklessly making me ready to do what in my own proper, natural heart, I durst not so much as dare? Is Ahab, Ahab? Is it I, God, or who, that lifts this arm? But if the great sun move not of himself; but is as an errand-boy in heaven; nor one single star can revolve, but by some invisible power; how then can this one small heart beat; this one small brain think thoughts; unless God does that beating, does that thinking, does that living, and not I. By heaven, man, we are turned round and round in this world, like yonder windlass, and Fate is the handspike. And all the time, lo! that smiling sky, and this unsounded sea! Look! see yon Albicore! who put it into him to chase and fang that flying-fish? Where do murderers go, man! Who's to doom, when the judge himself is dragged to the bar? But it is a mild, mild wind, and a mild looking sky; and the air smells now, as if it blew from a far-away meadow; they have been making hay somewhere under the slopes of the Andes, Starbuck, and the mowers are sleeping among the new-mown hay. Sleeping? Aye, toil we how we may, we all sleep at last on the field. Sleep? Aye, and rust amid greenness; as last year's scythes flung down, and left in the half-cut swaths—Starbuck!"

But blanched to a corpse's hue with despair, the Mate had stolen away.

Ahab crossed the deck to gaze over on the other side; but started at two reflected, fixed eyes in the water there. Fedallah was motionlessly leaning over the same rail.

Bark Greyhound *under sail, circa 1918.* — *Arthur F. Packard photograph.*

Spinner Collection

Chapter CXXXIII

The Chase—First Day

That night, in the mid-watch, when the old man—as his wont at intervals—stepped forth from the scuttle in which he leaned, and went to his pivot-hole, he suddenly thrust out his face fiercely, snuffing up the sea air as a sagacious ship's dog will, in drawing nigh to some barbarous isle. He declared that a whale must be near, and then ascertaining the precise bearing of the odor as nearly as possible, Ahab rapidly ordered the ship's course to be slightly altered, and the sail to be shortened.

The acute policy dictating these movements was sufficiently vindicated at daybreak, by the sight of a long sleek on the sea directly and lengthwise ahead, smooth as oil, and resembling in the pleated watery wrinkles bordering it, the polished metallic-like marks of some swift tide-rip, at the mouth of a deep, rapid stream.

"What d'ye see?" cried Ahab, flattening his face to the sky.

"Nothing, nothing, sir!" was the sound hailing down in reply.

"T'gallant sails!—stunsails! alow and aloft, and on both sides!"

All sail being set, he now cast loose the life-line, reserved for swaying him to the main royal-mast head; and in a few moments they were hoisting him thither, when, while but two thirds of the way aloft, and while peering ahead through the horizontal vacancy between the main-top-sail and top-gallant-sail, he raised a gull-like cry in the air, "There she blows!—there she blows! A hump like a snow-hill! It is Moby Dick!"

Fired by the cry which seemed simultaneously taken up by the three look-outs, the men on deck rushed to the rigging to behold the famous whale they had so long been pursuing. Ahab had now gained his final perch, some feet above the other look-outs, Tashtego standing just beneath him on the cap of the top-gallant mast, so that the Indian's head was almost on a level with Ahab's heel. From this height the whale was now seen some mile or so ahead, at every roll of the sea revealing his high sparkling hump, and regularly jetting his silent spout into the air. To the credulous mariners it seemed the same silent spout they had so long ago beheld in the moonlit Atlantic and Indian Oceans.

"There she blows! there she blows!—there she blows! There again!—there again!" he cried, in long-drawn, lingering, methodic tones, attuned to the gradual prolongings of the whale's visible jets. "He's going to

Masthead man aboard *Sunbeam*, 1904.
— Clifford W. Ashley photograph, New Bedford Whaling Museum.

sound! In stunsails! Down top-gallant-sails! Stand by three boats. Mr. Starbuck, remember, stay on board, and keep the ship. Helm there! Luff, luff a point! So; steady, man, steady! There go flukes! No, no; only black water! All ready the boats there? Stand by, stand by! Lower me, Mr. Starbuck; lower, lower,—quick, quicker!" and he slid through the air to the deck.

"He is heading straight to leeward, Sir," cried Stubb, 'right away from us; cannot have seen the ship yet."

"Be dumb, man! Stand by the braces! Hard down the helm!—brace up! Shiver her!—shiver her! So; well that! Boats, boats!"

Soon all the boats but Starbuck's were dropped; all the boat-sails set—all the paddles plying; with rippling swiftness, shooting to leeward; and Ahab heading the onset. A pale, death-glimmer lit up Fedallah's sunken eyes; a hideous motion gnawed his mouth.

Like noiseless nautilus shells, their light prows sped through the sea; but only slowly they neared the foe. As they neared him, the ocean grew still more smooth; seemed drawing a carpet over its waves; seemed a noon-meadow, so serenely it spread. At length the breathless hunter came so nigh his seemingly unsuspecting prey, that his entire dazzling hump was distinctly visible, sliding along the sea as if an isolated thing, and continually set in a revolving ring of finest, fleecy, greenish foam. He saw the vast, involved wrinkles of the slightly projecting head beyond. Before it, far out on the soft Turkish-rugged waters, went the glistening white shadow from his broad, milky forehead, a musical rippling playfully accompanying the shade; and behind, the blue waters interchangeably flowed over into the moving valley of his steady wake; and on either hand bright bubbles arose and danced by his side. But these were broken again by the light toes of hundreds of gay fowl softly feathering the sea, alternate with their fitful flight; and like to some flag-staff rising from the painted hull of an argosy, the tall but shattered pole of a recent lance projected from the white whale's back; and at intervals one of the cloud of soft-toed fowls hovering, and to and fro skimming like a canopy over the fish, silently perched and rocked on this pole, the long tail feathers streaming like pennons.

A gentle joyousness—a mighty mildness of repose in swiftness, invested the gliding whale. Not Jove, not that great majesty Supreme! Did surpass the glorified White Whale as he so divinely swam.

And thus, through the serene tranquillities of the tropical sea, among waves whose hand-clappings were suspended by exceeding rapture, Moby Dick moved on, still withholding from sight the full terrors of his submerged trunk, entirely hiding the wrenched hideousness of his jaw. But soon the fore part of him slowly rose from the water; for an instant his whole marbleized body formed a high arch, and warningly waving his bannered flukes in the air, the grand god revealed himself, sounded, and went out of sight.

With oars apeak, and paddles down, the sheets of their sails adrift, the three boats now stilly floated, awaiting Moby Dick's reappearance. "An hour," said Ahab, standing rooted in his boat's stern; and he gazed beyond the whale's place, towards the dim blue spaces and wide wooing vacancies to leeward. It was only an instant; for again his eyes seemed whirling round in his head as he swept the watery circle. The breeze now freshened; the sea began to swell.

But suddenly as he peered down and down into its depths, he profoundly saw a white living spot no bigger than a white weasel, with wonderful celerity uprising, and magnifying as it rose, till it turned, and then there were plainly revealed two long crooked rows of white, glistening teeth, floating up from the undiscoverable bottom. It was Moby Dick's open mouth and scrolled jaw; his vast, shadowed bulk still half blending with the blue of the sea. The glittering mouth yawned beneath the boat like an open-doored marble tomb; and giving one side-long sweep with his steering oar, Ahab whirled the craft aside from this tremendous apparition.

But as if perceiving this strategem, Moby Dick, with that malicious intelligence ascribed to him, sidelingly transplanted himself, as it were, in an instant, shooting his pleated head lengthwise beneath the boat.

Through and through; through every plank and each rib, it thrilled for an instant, the whale obliquely lying on his back, in the manner of a biting shark, slowly and feelingly taking its bows full within his mouth, so that the long, narrow, scrolled lower jaw curled high up into the open air, and one of the teeth caught in a rowlock. The bluish pearl-white of the inside of the jaw was within six inches of Ahab's head, and reached higher than that. In this attitude the White Whale now shook the slight cedar as a mildly cruel cat her mouse. With unastonished eyes Fedallah gazed, and crossed his arms; but the tiger-yellow crew were tumbling over each other's heads to gain the uttermost stern.

And now, while both elastic gunwales were springing in and out, as the whale dallied with the doomed craft in this devilish way; and from his body being submerged beneath the boat, he could not be darted at from the bows, for the bows were almost inside of him, as it were; and while the other boats involuntarily paused, as before a quick crisis impossible to withstand, then it was that monomaniac Ahab, furious with this tantalizing vicinity of his foe, which placed him all alive and helpless in the very jaws he hated; frenzied with all this, he seized the long bone with his naked hands, and wildly strove to wrench it from its gripe. As now he thus vainly strove, the jaw slipped from him; the frail gunwales bent in, collapsed, and snapped, as both jaws, like an enormous shears, sliding further aft, bit the craft completely

"Whale in a fury." —Illustration by Rockwell Kent.

©1930 by R. R. Donnelley & Sons Co., with permission

in twain, and locked themselves fast again in the sea, midway between the two floating wrecks. These floated aside, the broken ends drooping, the crew at the stern-wreck clinging to the gunwales, and striving to hold fast to the oars to lash them across.

At that preluding moment, ere the boat was yet snapped, Ahab, the first to perceive the whale's intent, by the crafty upraising of his head, a movement that loosed his hold for the time; at that moment his hand had made one final effort to push the boat out of the bite. But only slipping further into the whale's mouth, and tilting over sideways as it slipped, the boat had shaken off his hold on the jaw; spilled him out of it, as he leaned to the push; and so he fell flat-faced upon the sea.

Ripplingly withdrawing from his prey, Moby Dick now lay at a little distance, vertically thrusting his oblong white head up and down in the billows; and at the same time slowly revolving his whole spindled body; so that when his vast wrinkled forehead rose—some twenty or more feet out of the water—the now rising swells, with all their confluent waves, dazzlingly broke against it; vindictively tossing their shivered spray still higher into the air.

But soon resuming his horizontal attitude, Moby Dick swam swiftly round and round the wrecked crew; sideways churning the water in his vengeful wake, as if lashing himself up to still another and more deadly

"Thrown into a whale's mouth," **1886.** — *Engraving by W. Roberts, from Delano: Wanderings....*

Stove Boat, circa 1840. *This watercolor is one of four drawings on the same theme, done by an unknown whaleman, possibly aboard the ship* Young Phoenix *of New Bedford.* — *Source: Kugler. Companion art on page 109.*

New Bedford Whaling Museum

"A Stove Boat," 1874. — *From Davis:* Nimrod of the Sea.

assault. Meanwhile Ahab half smothered in the foam of the whale's insolent tail, and too much of a cripple to swim,—though he could still keep afloat, even in the heart of such a whirlpool as that; helpless Ahab's head was seen, like a tossed bubble which the least chance shock might burst. From the boat's fragmentary stern, Fedallah incuriously and mildly eyed him; the clinging crew, at the other drifting end, could not succor him; more than enough was it for them to look to themselves. For so revolvingly appalling was the White Whale's aspect, and so planetarily swift the ever-contracting circles he made, that he seemed horizontally swooping upon them. And though the other boats, unharmed, still hovered hard by; still they dared not pull into the eddy to strike, lest that should be the signal for the instant destruction of the jeopardized castaways, Ahab and all; nor in that case could they themselves hope to escape. With straining eyes, then, they remained on the outer edge of the direful zone, whose centre had now become the old man's head.

Meantime, from the beginning all this had been descried from the ship's mast heads; and squaring her yards, she had borne down upon the scene; and was now so nigh, that Ahab in the water hailed her,—"Sail on the whale!—Drive him off!"

The Pequod's prows were pointed; and breaking up the charmed circle, she effectually parted the White Whale from his victim. As he sullenly swam off, the boats flew to the rescue.

Dragged into Stubb's boat with blood-shot, blinded eyes, the white brine caking in his wrinkles; the long tension of Ahab's bodily strength did crack, and helplessly he yielded to his body's doom: for a time, lying all crushed in the bottom of Stubb's boat, like one trodden under foot of herds of elephants.

But this intensity of his physical prostration did but so much the more abbreviate it.

"The harpoon," said Ahab, half way rising, and draggingly leaning on one bended arm—"is it safe?"

"Aye, sir, for it was not darted; this is it," said Stubb, showing it.

"Lay it before me;—any missing men?"

"One, two, three, four, five;—there were five oars, Sir, and here are five men."

"That's good.—Help me, man; I wish to stand. So, so, I see him! there! there! going to leeward still; what a leaping spout! Hands off from me! The eternal sap runs up in Ahab's bones again! Set the sail; out oars; the helm!"

It is often the case that when a boat is stove, its crew, being picked up by another boat, help to work that second boat; and the chase is thus continued with what is called double-banked oars. It was thus now. But the added power of the boat did not equal the added power of the whale, for he seemed to have treble-banked his every fin; swimming with a velocity which plainly showed, that if now, under these circumstances, pushed on, the chase would prove an indefinitely prolonged, if not a hopeless one. The ship itself, then, as it sometimes happens, offered the most promising intermediate means of overtaking the chase. Accordingly, the boats now made for her, and were soon swayed up to their cranes—the two parts of the wrecked boat having been previously secured by her—and then hoisting everything to her side, and stacking her canvas high up, and sideways outstretching it with stun-sails; the Pequod bore down in the leeward wake of Moby Dick.

At the well known, methodic intervals, the whale's glittering spout was announced from the manned mastheads; and when he would be reported as just gone down, Ahab would take the time, and then pacing the deck, binnacle-watch in hand, so soon as the last second of the allotted hour expired, his voice was heard.— "D'ye see him?" and if the reply was, "No, sir!" straightway he commanded them to lift him to his perch. In this way the day wore on; Ahab, now aloft and motionless; anon, unrestingly pacing the planks.

As he was thus walking, uttering no sound, except to hail the men aloft, or to bid them hoist a sail still higher, or to spread one to a still greater breadth—thus to and fro pacing, beneath his slouched hat, at every turn he passed his own wrecked boat, which had been dropped upon the quarter-deck, and lay there reversed; broken bow to shattered stern. At last he paused before it; and as in an already over-clouded sky fresh troops of clouds will sometimes sail across, so over the old man's face there now stole some such added gloom as this.

Stubb saw him pause; and perhaps intending, not vainly, though, to evince his own unabated fortitude, and thus keep up a valiant place in his Captain's mind, he advanced, and eyeing the wreck exclaimed—"The thistle the ass refused; it pricked his mouth too keenly, sir; ha! ha!"

"What soulless thing is this that laughs before a wreck? Man, man! Groan nor laugh should be heard before a wreck."

"Aye, sir," said Starbuck drawing near, "'tis a solemn sight; an omen, and an ill one."

"Omen? omen?—the dictionary! If the gods think to speak outright to man, they will honorably speak outright; not shake their heads, and give an old wives' darkling hint.—Begone! Ye two are the opposite poles of one thing; Starbuck is Stubb reversed, and Stubb is Starbuck; and ye two are all mankind; and Ahab stands alone among the millions of the peopled earth, nor gods nor men his neighbors! Aloft there! D'ye see him? Sing out for every spout, though he spout ten times a second!"

The day was nearly done; only the hem of his golden robe was rustling. Soon, it was almost dark, but the lookout men still remained unset.

"Can't see the spout now, sir;—too dark"—cried a voice from the air.

"How heading when last seen?"

"As before, sir,—straight to leeward."

"Good! he will travel slower now 'tis night. Down royals and top-gallant stun-sails, Mr. Starbuck. We must not run over him before morning; he's making a passage now, and may heave-to a while. Helm there! keep her full before the wind!—Aloft! come down!—Mr. Stubb, send a fresh hand to the fore-mast head, and see it manned till morning."

And so saying, he placed himself half way within the scuttle, and slouching his hat, stood there till dawn, except when at intervals rousing himself to see how the night wore on.

"The Chase," 2001. "On each soft side—coincident with the parted swell, that but once leaving him, then flowed so wide away—on each bright side, the whale shed off enticings. No wonder there had been some among the hunters who namelessly transported and allured by all this serenity, had ventured to assail it; but had fatally found that quietude but the vesture of tornadoes. Yet calm, enticing calm, oh, whale! thou glidest on, to all who for the first time eye thee, no matter how many in that same way thou may'st have bejuggled and destroyed before." — Chapter 133. Oil painting by Arthur Moniz.

Chapter CXXXIV

The Chase—Second Day

"There she breaches!" — *From* Moby-Dick, *published by Charles Scribners and Sons, 1899.*

At day-break, the three mast-heads were punctually manned afresh.

"D'ye see him?" cried Ahab, after allowing a little space for the light to spread.

"See nothing, Sir."

"Turn up all hands and make sail! he travels faster than I thought for;—the top-gallant sails!—aye, they should have been kept on her all night. But no matter—'tis but resting for the rush."

Here be it said, that this pertinacious pursuit of one particular whale, continued through day into night, and through night into day, is a thing by no means unprecedented in the South sea fishery. For such is the wonderful skill, prescience of experience, and invincible confidence acquired by some great natural geniuses among the Nantucket commanders; that from the simple observation of a whale when last descried, they will, under certain given circumstances, pretty accurately foretell both the direction in which he will continue to swim for a time, while out of sight, as well as his probable rate of progression during that period. So that to this hunter's wondrous skill, the proverbial evanescence of a thing writ in water, a wake, is to all desired purposes well nigh as reliable as the steadfast land. There are occasions when these Nantucketers time that other Leviathan of the deep, according to the observed humor of his speed; and say to themselves, so many hours hence this whale will have gone two hundred miles, will have about reached this or that degree of latitude or longitude. But to render this acuteness at all successful in the end, the wind and the sea must be the whaleman's allies; for of what present avail to the becalmed or windbound mariner is the skill that assures him he is exactly ninety-three leagues and a quarter from his port?

The ship tore on; leaving such a furrow in the sea as when a cannon-ball, missent, becomes a plough-share and turns up the level field.

"There she blows—she blows!—she blows!—right ahead!" was now the mast-head cry.

"Aye, aye!" cried Stubb. "I knew it—ye can't escape—blow on and split your spout, O whale! the mad fiend himself is after ye! blow your trump—blister your lungs!—Ahab will dam off your blood, as a miller shuts his water-gate upon the stream!"

And Stubb did but speak out for well nigh all that crew. The frenzies of the chase had by this time worked them bubblingly up, like old wine worked anew. Whatever pale fears and forebodings some of them might have felt before; these were not only now kept out of sight through the growing awe of Ahab, but they were broken up, and on all sides routed, as timid prairie hares that scatter before the bounding bison. The wind that made great bellies of their sails, and rushed the vessel on by arms invisible as irresistible; this seemed the symbol of that unseen agency which so enslaved them to the race.

They were one man, not thirty. For as the one ship that held them all; though it was put together of all contrasting things—oak, and maple, and pine wood; iron, and pitch, and hemp—yet all these ran into each other in the one concrete hull, which shot on its way, both balanced and directed by the long central keel; even so, all the individualities of the crew, this man's valor, that man's fear; guilt and guiltiness, all varieties were welded into oneness, and were all directed to that fatal goal which Ahab their one lord and keel did point to.

The rigging lived. The mast-heads, like the tops of tall palms, were outspreadingly tufted with arms and legs. Clinging to a spar with one hand, some reached forth the other with impatient wavings; others, shading their eyes from the vivid sunlight, sat far out on the rocking yards; all the spars in full bearing of mortals, ready and ripe for their fate. Ah! how they still strove through that infinite blueness to seek out the thing that might destroy them!

"Why sing ye not out for him, if ye see him?" cried Ahab, when, after the lapse of some minutes since the first cry, no more had been heard. "Sway me up, men; ye have been deceived; not Moby Dick casts one odd jet that way, and then disappears."

It was even so; in their headlong eagerness, the men had mistaken some other thing for the whale-spout, as the event itself soon proved; for hardly had Ahab reached his perch; hardly was the rope belayed to its pin on deck, when he struck the key-note to an orchestra, that made the air vibrate as with the combined discharges of rifles. The triumphant halloo of thirty buckskin lungs was heard, as—much nearer to the ship than the place of the imaginary jet, less than a mile ahead—Moby Dick bodily burst into view! For not by any calm and indolent spoutings; not by the peaceable gush of that mystic fountain in his head, did the White Whale now reveal his vicinity; but by the far more wondrous phenomenon of breaching. Rising with his utmost velocity from the furthest depths, the Sperm Whale thus booms his entire bulk into the pure element of air, and piling up a mountain of dazzling foam, shows his place to the distance of seven miles and more. In those moments, the torn, enraged waves he shakes off, seem his mane; in some cases, this breaching is his act of defiance.

"There she breaches! there she breaches!" was the cry.

"Aye, breach your last to the sun, Moby Dick!" cried Ahab, "thy hour and thy harpoon are at hand!—Down! down all of ye, but one man at the fore. The boats!—stand by!"

Unmindful of the tedious rope-ladders of the shrouds, the men, like shooting stars, slid to the deck, by the isolated back-stays and halyards; while Ahab, less dartingly, but still rapidly was dropped from his perch.

"Lower away," he cried, so soon as he had reached his boat—a spare one, rigged the afternoon previous. "Mr. Starbuck, the ship is thine—keep away from the boats, but keep near them. Lower, all!"

As if to strike a quick terror into them, by this time being the first assailant himself, Moby Dick had turned, and was now coming for the three crews. Ahab's boat was central; and cheering his men, he told them he would take the whale head-and-head,—that is, pull straight up to his forehead,—a not uncommon thing; for when within a certain limit, such a course excludes the coming onset from the whale's sidelong vision. But ere that close limit was gained, and while yet all three boats were plain as the ship's three masts to his eye; the White Whale churning himself into furious speed, almost in an instant as it were, rushing among the boats with open jaws, and a lashing tail, offered appalling battle on every side; and heedless of the irons darted at him from every boat, seemed only intent on annihilating each separate plank of which those boats were made. But skilfully manoeuvred, incessantly wheeling like trained chargers in the field; the boats for a while eluded him; though, at times, but by a plank's breadth; while all the time, Ahab's unearthly slogan tore every other cry but his to shreds.

But at last in his untraceable evolutions, the White Whale so crossed and recrossed, and in a thousand ways entangled the slack of the three lines now fast to him, that they foreshortened, and, of themselves, warped the devoted boats towards the planted irons in him; though now for a moment the whale drew aside a little, as if to rally for a more tremendous charge. Seizing that opportunity, Ahab first paid out more line: and then was rapidly hauling and jerking in upon it again—hoping that way to disencumber it of some snarls—when lo!—a sight more savage than the embattled teeth of sharks!

Caught and twisted—corkscrewed in the mazes of the line, loose harpoons and lances, with all their bristling barbs and points, came flashing and dripping up to the chocks in the bows of Ahab's boat. Only one

"The Flurry," 1859. Not long after Moby-Dick *was published, a story entitled "Whaling Adventures," accompanied by four engravings, appeared in* Frank Leslie's Illustrated Newspaper. *The writer, Peter L. Dumont, like Melville, was a young greenhand from New York state. He told of an aggressive sperm whale waging battle with his predators. "His tail [was] extended in the air and striking out in every direction, as if bent on revenge for the injuries we had inflicted on him. He came rapidly onward towards the captain's boat, throwing huge sheets of water on the crew at every sweep of his gigantic tail, and filling it to the very brim."* — Engraving (digitally altered) by Granville Perkins. Source: Ingalls.

"Stove Boat," 1846.
— Ilustration by Alonzo Hartwell. From Delano: Wanderings and Adventures....

"The Perils of the Whale Fishery," 1840.
— From a German travel narrative. Kendall Collection, NBWM.

thing could be done. Seizing the boat-knife, he critically reached within—through—and then, without—the rays of steel; dragged in the line beyond, passed it, inboard, to the bowsman, and then, twice sundering the rope near the chocks—dropped the intercepted fagot of steel into the sea; and was all fast again. That instant, the White Whale made a sudden rush among the remaining tangles of the other lines; by so doing, irresistibly dragged the more involved boats of Stubb and Flask towards his flukes; dashed them together like two rolling husks on a surf-beaten beach, and then, diving down into the sea, disappeared in a boiling maelstrom, in which, for a space, the odorous cedar chips of the wrecks danced round and round.

While the two crews were yet circling in the waters, reaching out after the revolving line-tubs, oars, and other floating furniture, while aslope little Flask bobbed up and down like an empty vial, twitching his legs upwards to escape the dreaded jaws of sharks; and Stubb was lustily singing out for some one to ladle him up; and while the old man's line—now parting—admitted of his pulling into the creamy pool to rescue whom he could;—in that wild simultaneousness of a thousand concreted perils,—Ahab's yet unstricken boat seemed drawn up towards Heaven by invisible wires,—as, arrow-like, shooting perpendicularly from the sea, the White Whale dashed his broad forehead against its bottom, and sent it, turning over and over, into the air; till it fell again—gunwale downwards—and Ahab and his men struggled out from under it, like seals from a seaside cave.

The first uprising momentum of the whale—modifying its direction as he struck the surface— involuntarily launched him along it, to a little distance from the centre of the destruction he had made; and with his back to it, he now lay for a moment slowly feeling with his flukes from side to side; and whenever a stray oar, bit of plank, the least chip or crumb of the boats touched his skin, his tail swiftly drew back, and came sideways smiting the sea. But soon, as if satisfied that his work for that time was done, he pushed his pleated forehead through the ocean, and trailing after him the intertangled lines, continued his leeward way at a traveller's methodic pace.

As before, the attentive ship having descried the whole fight, again came bearing down to the rescue, and dropping a boat, picked up the floating mariners, tubs, oars and whatever else could be caught at, and safely landed them on her decks. Some sprained shoulders, wrists, and ankles; livid contusions; wrenched harpoons and lances; inextricable intricacies of rope; shattered oars and planks; all these were there; but no fatal or even serious ill seemed to have befallen any one. As with Fedallah the day before, so Ahab was now found grimly clinging to his boat's broken half, which afforded a comparatively easy float; nor did it so exhaust him as the previous day's mishap.

But when he was helped to the deck, all eyes were fastened upon him; as instead of standing by himself he still half-hung upon the shoulder of Starbuck, who had thus far been the foremost to assist him. His ivory leg had been snapped off, leaving but one short sharp splinter.

"Aye, aye," Starbuck, "'tis sweet to lean sometimes, be the leaner who he will; and would old Ahab had leaned oftener than he has."

"The ferrule has not stood, sir," said the carpenter, now coming up; "I put good work into that leg."

"But no bones broken, sir, I hope," said Stubb with true concern.

"Aye! and all splintered to pieces, Stubb!—d'ye see it.—But even with a broken bone, old Ahab is untouched; and I account no living bone of mine one jot more me, than this dead one that's lost. Nor White Whale, nor man, nor fiend, can so much as graze old Ahab in his own proper and inaccessible being.—Aloft there! which way?"

"Dead to leeward, sir."

"Up helm, then; pile on the sail again, ship keepers! down the rest of the spare boats and rig them—Mr. Starbuck away, and muster the boat's crews."

"Let me first help thee towards the bulwarks, sir."

"Oh, oh, oh! how this splinter gores me now! Accursed fate! that the unconquerable captain in the soul should have such a craven mate!"

"Sir?"

"My body, man, not thee. Give me something for a cane—there, that shivered lance will do. Muster the men. Surely I have not seen him yet. By heaven it cannot be!—missing?—quick! call them all."

The old man's hinted thought was true. Upon mustering the company, the Parsee was not there.

"The Parsee!" cried Stubb—"he must have been caught in—"

"The black vomit wrench thee!—run all of ye above, alow, cabin, forecastle—find him—not gone—not gone!"

But quickly they returned to him with the tidings that the Parsee was nowhere to be found.

"Aye, sir," said Stubb— "caught among the tangles of your line—I thought I saw him dragging under."

"My line! my line? Gone?—gone? What means that little word?—What death-knell rings in it, that old Ahab shakes as if he were the belfry. The harpoon, too!—toss over the litter there,—d'ye see it?—the forged iron, men, the White Whale's—no, no, no,—blistered fool; this hand did dart it!—'tis in the fish!—Aloft there! keep him nailed—quick!—all hands to the rigging of the boats—collect the oars—harpooneers! the irons, the irons!—hoist the royals higher—a pull on all the sheets!—helm there! steady, steady for your life! I'll ten times girdle the unmeasured globe; yea and dive straight through it, but I'll slay him yet!"

"Great God! but for one single instant show thyself," cried Starbuck; "never, never wilt thou capture him, old man—In Jesus' name no more of this, that's worse than devil's madness. Two days chased; twice stove to splinters; thy very leg once more snatched from under thee; thy evil shadow gone—all good angels mobbing thee with warnings:—what more wouldst thou have?—Shall we keep chasing this murderous fish till he swamps the last man? Shall we be dragged by him to the bottom of the sea? Oh, oh,—Impiety and blasphemy to hunt him more!"

"Starbuck, of late I've felt strangely moved to thee; ever since that hour we both saw—thou know'st what, in one another's eyes. But in this matter of the whale, be the front of thy face to me as the palm of this hand—a lipless, unfeatured blank. Ahab is for ever Ahab, man. This whole act's immutably decreed. 'Twas rehearsed by thee and me a billion years before this ocean rolled. Fool! I am the Fates' lieutenant; I act under orders. Look thou,

Capsized, 1843. "…*The whale up flukes and capsized the boat in the same instant and thus they were…thrown in the sea without a moments notice.*" — *From the journal of Joseph Bogart Hershey, Schooner* Esquimaux *of Provincetown.*

underling! that thou obeyest mine.—Stand round me, men. Ye see an old man cut down to the stump; leaning on a shivered lance; propped up on a lonely foot. 'Tis Ahab—his body's part; but Ahab's soul's a centipede, that moves upon a hundred legs. I feel strained, half stranded, as ropes that tow dismasted frigates in a gale; and I may look so. But ere I break, ye'll hear me crack; and till ye hear *that*, know that Ahab's hawser tows his purpose yet. Believe ye, men, in the things called omens? Then laugh aloud, and cry encore! For ere they drown, drowning things will twice rise to the surface; then rise again, to sink for evermore. So with Moby Dick—two days he's floated—to-morrow will be the third. Aye, men, he'll rise once more,—but only to spout his last! D'ye feel brave men, brave?"

"As fearless fire," cried Stubb.

"And as mechanical," muttered Ahab. Then as the men went forward, he muttered on:—"The things called omens! And yesterday I talked the same to Starbuck there, concerning my broken boat. Oh! how valiantly I seek to drive out of others' hearts what's clinched so fast in mine!—The Parsee—the Parsee!—gone, gone? and he was to go before:—but still was to be seen again ere I could perish—How's that?—There's a riddle now might baffle all the lawyers backed by the ghosts of the whole line of judges:—like a hawk's beak it pecks my brain. *I'll, I'll solve it, though!*"

When dusk descended, the whale was still in sight to leeward.

So once more the sail was shortened, and everything passed nearly as on the previous night; only, the sound of hammers, and the hum of the grindstone was heard till nearly daylight, as the men toiled by lanterns in the complete and careful rigging of the spare boats and sharpening their fresh weapons for the morrow. Meantime, of the broken keel of Ahab's wrecked craft the carpenter made him another leg; while still as on the night before, slouched Ahab stood fixed within his scuttle; his hid, heliotrope glance anticipatingly gone backward on its dial; sat due eastward for the earliest sun.

Chapter CXXXV

The Chase—Third Day

Ship *Niger* of New Bedford, circa 1880. — *William Bradford photograph*

The morning of the third day dawned fair and fresh, and once more the solitary night-man at the fore-mast-head was relieved by crowds of the daylight look-outs, who dotted every mast and almost every spar.

"D'ye see him?" cried Ahab; but the whale was not yet in sight.

"In his infallible wake, though; but follow that wake, that's all. Helm there; steady, as thou goest, and hast been going. What a lovely day again! were it a new-made world, and made for a summer-house to the angels, and this morning the first of its throwing open to them, a fairer day could not dawn upon that world. Here's food for thought, had Ahab time to think; but Ahab never thinks; he only feels, feels, feels; *that's* tingling enough for mortal man! to think's audacity. God only has that right and privilege. Were I the wind, I'd blow no more on such a wicked, miserable world. I'd crawl somewhere to a cave, and slink there. And yet, 'tis a noble and heroic thing, the wind! who ever conquered it? Run tilting at it, and you but run through it. Ha! a coward wind that strikes stark naked men, but will not stand to receive a single blow. And yet, I say again, and swear it now, that there's something all glorious and gracious in the wind. And by the eternal Poles! these same Trades that so directly blow my good ship on; these Trades, or something like them—something so unchangeable, and full as strong, blow my keeled soul along! To it! Aloft there! What d'ye see?"

"Nothing, sir."

"Nothing! and noon at hand! The doubloon goes a-begging! See the sun! Aye, aye, it must be so. I've over-sailed him. How, got the start? Aye, he's chasing *me* now;

Ship James Arnold at sea, 1893. "...How the wild winds blow it (hair); they whip it about me as the torn shreds of split sails lash the tossed ship they cling to. A vile wind that has no doubt blown ere this through prison corridors and cells, and wards of hospitals, and ventilated them, and now comes blowing hither as innocent as fleeces. Out upon it!—it's tainted." — *Ahab: Chapter 135.*

William R. Hegarty Collection

not I, him—that's bad; I might have known it, too. Fool! the lines—the harpoons he's towing. Aye, aye, I have run him by last night. About! about! Come down, all of ye, but the regular look outs! Man the braces!"

Steering as she had done, the wind had been somewhat on the Pequod's quarter, so that now being pointed in the reverse direction, the braced ship sailed hard upon the breeze as she rechurned the cream in her own white wake.

"Against the wind he now steers for the open jaw," murmured Starbuck to himself, as he coiled the new-hauled main-brace upon the rail. "God keep us, but already my bones feel damp within me, and from the inside wet my flesh. I misdoubt me that I disobey my God in obeying him!"

"Stand by to sway me up!" cried Ahab, advancing to the hempen basket. "We should meet him soon."

"Aye, aye, sir," and straightway Starbuck did Ahab's bidding, and once more Ahab swung on high.

A whole hour now passed; gold-beaten out to ages. Time itself now held long breaths with keen suspense. But at last, some three points off the weather bow, Ahab descried the spout again, and instantly from the three mast-heads three shrieks went up as if the tongues of fire had voiced it.

"Forehead to forehead I meet thee, this third time, Moby Dick! On deck there!—brace sharper up; crowd her into the wind's eye. He's too far off to lower yet, Mr. Starbuck! So, so; he travels fast, and I must down. But let me have one more good round look aloft here at the sea; there's time for that. An old, old sight, and yet somehow so young; aye, and not changed a wink since I first saw it, a boy, from the sand-hills of Nantucket! The same!—the same!—the same to Noah as to me. There's a soft shower to leeward. Such lovely leeward-ings! Leeward! the White Whale goes that way; look to windward, then; the better if the bitterer quarter. But good bye, good bye, old mast-head! What's this?—green? aye, tiny mosses in these warped cracks. No such green weather stains on Ahab's head! There's the difference now between man's old age and matter's. But aye, old mast, we both grow old together; sound in our hulls, though, are we not, my ship? Aye, minus a leg, that's all. What's that he said? he should still go before me, my pilot; and yet to be seen again? But where? Will I have eyes at the bottom of the sea, supposing I descend those endless stairs? and all night I've been sailing from him, wherever he did sink to. Aye, aye, like many more thou told'st direful truth as touching thyself, O Parsee; but, Ahab, there thy shot fell short. Good by, mast-head—keep a good eye upon the whale, the while I'm gone. We'll talk to-morrow, nay, to-night, when the White Whale lies down there, tied by head and tail."

He gave the word; and still gazing round him, was steadily lowered through the cloven blue air to the deck.

In due time the boats were lowered, but as standing in his shallop's stern, Ahab just hovered upon the point of the descent, he waved to the mate,—who held one of the tackle-ropes on deck—and bade him pause.

"Starbuck!"

"Sir?"

"For the third time my soul's ship starts upon this voyage, Starbuck."

"Aye, sir, thou wilt have it so."

"Some ships sail from their ports, and ever afterwards are missing, Starbuck!"

"Truth, Sir: saddest truth."

"Some men die at ebb tide; some at low water; some at the full of the flood;—and I feel now like a billow that's all one crested comb, Starbuck. I am old;—shake hands with me, man."

Their hands met; their eyes fastened; Starbuck's tears the glue.

"Oh, my captain, my captain!—noble heart—go not—go not!—see, it's a brave man that weeps; how great the agony of the persuasion then!"

"Lower away!"—cried Ahab, tossing the mate's arm from him. "Stand by the crew!"

In an instant the boat was pulling round close under the stern.

"The sharks! the sharks!" cried a voice from the low cabin-window there; "O master, my master, come back!"

But Ahab heard nothing; for his own voice was high-lifted then; and the boat leaped on.

Yet the voice spake true; for scarce had he pushed from the ship, when numbers of sharks, seemingly rising from out the dark waters beneath the hull, maliciously snapped at the blades of the oars, every time they dipped

Lowering away, circa 1857. *The ship* Chili *of New Bedford lowers for sperm whales in the Indian Ocean.* — *From the journal of Rodolphus W. Dexter.*

Kendall Collection, NBWM

in the water; and in this way accompanied the boat with their bites. It is a thing not uncommonly happening to the whale-boats in those swarming seas; the sharks at times apparently following them in the same prescient way that vultures hover over the banners of marching regiments in the east. But these were the first sharks that had been observed by the Pequod since the White Whale had been first descried; and whether it was that Ahab's crew were all such tiger-yellow barbarians, and therefore their flesh more musky to the senses of the sharks—a matter sometimes well known to affect them,—however it was, they seemed to follow that one boat without molesting the others.

"Heart of wrought steel!" murmured Starbuck gazing over the side, and following with his eyes the receding boat—"canst thou yet ring boldly to that sight?—lowering thy keel among ravening sharks, and followed by them, open-mouthed to the chase; and this the critical third day?—For when three days flow together in one continuous intense pursuit; be sure the first is the morning, the second the noon, and the third the evening and the end of that thing—be that end what it may. Oh! my God! what is this that shoots through me, and leaves me so deadly calm, yet expectant,—fixed at the top of a shudder! Future things swim before me, as in empty outlines and skeletons; all the past is somehow grown dim. Is my journey's end coming? My legs feel faint; like his who has footed it all day. Feel thy heart,—beats it yet?—Stir thyself, Starbuck!—stave it off—move, move! speak aloud!—Mast-head there! See ye my boy's hand on the hill?—Crazed;—aloft there!—keep thy keenest eye upon the boats:—mark well the whale!—Ho! again!—drive off that hawk! see! he pecks—he tears the vane"—pointing to the red flag flying at the main-truck—"Ha! he soars away with it!—Where's the old man now? sees't thou that sight, oh Ahab!—shudder, shudder!"

The boats had not gone very far, when by a signal from the mast-heads—a downward pointed arm, Ahab knew that the whale had sounded; but intending to be near him at the next rising, he held on his way a little sideways from the vessel; the becharmed crew maintaining the profoundest silence, as the head-beat waves hammered and hammered against the opposing bow.

"Drive, drive in your nails, oh ye waves! to their uttermost heads, drive them in! ye but strike a thing without a lid; and no coffin and no hearse can be mine: —and hemp only can kill me! Ha! ha!"

Suddenly the waters around them slowly swelled in broad circles; then quickly upheaved, as if sideways sliding from a submerged berg of ice, swiftly rising to the surface. A low rumbling sound was heard; a subterrane-

"Moby Dick Transcendent," 1930. — *Illustration by Rockwell Kent.*
©1930 by R. R. Donnelley & Sons Co., with permission.

ous hum; and then all held their breaths; as bedraggled with trailing ropes, and harpoons, and lances, a vast form shot lengthwise, but obliquely from the sea. Shrouded in a thin drooping veil of mist, it hovered for a moment in the rainbowed air; and then fell swamping back into the deep. Crushed thirty feet upwards, the waters flashed for an instant like heaps of fountains, then brokenly sank in a shower of flakes, leaving the circling surface creamed like new milk round the marble trunk of the whale.

"Give way!" cried Ahab to the oarsmen, and the boats darted forward to the attack; but maddened by yesterday's fresh irons that corroded in him, Moby Dick seemed combinedly possessed by all the angels that fell from heaven. The wide tiers of welded tendons overspreading his broad white forehead, beneath the transparent skin, looked knitted together; as head on, he came churning his tail among the boats; and once more flailed them apart; spilling out the irons and lances from the two mates' boats, and dashing in one side of the upper part of their bows, but leaving Ahab's almost without a scar.

While Daggoo and Queequeg were stopping the strained planks; and as the whale swimming out from them, turned, and showed one entire flank as he shot by them again; at that moment a quick cry went up. Lashed round and round to the fish's back; pinioned in the turns upon turns in which, during the past night, the whale had reeled the involutions of the lines around him, the half torn body of the Parsee was seen; his sable raiment frayed to shreds; his distended eyes turned full upon old Ahab.

The harpoon dropped from his hand.

"Befooled, befooled!"—drawing in a long lean breath—"Aye, Parsee! I see thee again.—Aye, and thou goest before; and this, this then is the hearse that thou didst promise. But I hold thee to the last letter of thy

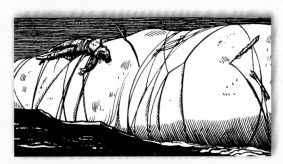

The hearse, 1930. — *Illustration by Rockwell Kent.*
©1930 by R. R. Donnelley & Sons Co., with permission.

word. Where is the second hearse? Away, mates, to the ship! those boats are useless now; repair them if ye can in time, and return to me; if not, Ahab is enough to die—Down, men! the first thing that but offers to jump from this boat I stand in, that thing I harpoon. Ye are not other men, but my arms and my legs; and so obey me.—Where's the whale? gone down again?"

But he looked too nigh the boat; for as if bent upon escaping with the corpse he bore, and as if the particular place of the last encounter had been but a stage in his leeward voyage, Moby Dick was now again steadily swimming forward; and had almost passed the ship,—which thus far had been sailing in the contrary direction to him, though for the present her headway had been stopped. He seemed swimming with his utmost velocity, and now only intent upon pursuing his own straight path in the sea.

"Oh! Ahab," cried Starbuck, "not too late is it, even now, the third day, to desist. See! Moby Dick seeks thee not. It is thou, thou, that madly seekest him!"

Setting sail to the rising wind, the lonely boat was swiftly impelled to leeward, by both oars and canvas. And at last when Ahab was sliding by the vessel, so near as plainly to distinguish Starbuck's face as he leaned over the rail, he hailed him to turn the vessel about, and follow him, not too swiftly, at a judicious interval. Glancing upwards, he saw Tashtego, Queequeg, and Daggoo, eagerly mounting to the three mast-heads; while the oarsmen were rocking in the two staved boats which had but just been hoisted to the side, and were busily at work in repairing them. One after the other, through the portholes, as he sped, he also caught flying glimpses of Stubb and Flask, busying themselves on deck among bundles of new irons and lances. As he saw all this; as he heard the hammers in the broken boats; far other hammers seemed driving a nail into his heart. But he rallied. And now marking that the vane or flag was gone from the main-mast-head, he shouted to Tashtego, who had just gained that perch, to descend again for another flag, and a hammer and nails, and so nail it to the mast.

Whether fagged by the three days' running chase, and the resistance to his swimming in the knotted hamper he bore; or whether it was some latent deceitfulness and malice in him: whichever was true, the White Whale's way now began to abate, as it seemed, from the boat so rapidly nearing him once more. And still as Ahab glided over the waves the unpitying sharks accompanied him; and so pertinaciously stuck to the boat; and so continually bit at the plying oars, that the blades became jagged and crunched, and left small splinters in the sea, at almost every dip.

"Heed them not! those teeth but give new rowlocks to your oars. Pull on! 'tis the better rest, the shark's jaw than the yielding water."

"But at every bite, Sir, the thin blades grow smaller and smaller!"

"They will last long enough! pull on!—But who can tell"—he muttered—"whether these sharks swim to feast on the whale or on Ahab?—But pull on! Aye, all alive, now—we near him. The helm! take the helm; let me pass,"—and so saying, two of the oarsmen helped him forward to the bows of the still flying boat.

At length as the craft was cast to one side, and ran ranging along with the White Whale's flank, he seemed strangely oblivious of its advance—as the whale sometimes will—and Ahab was fairly within the smoky mountain mist, which, thrown off from the whale's spout, curled round his great, Monadnock hump; he was even thus close to him; when, with body arched back, and both arms lengthwise high-lifted to the poise, he darted his fierce iron, and his far fiercer curse into the hated whale. As both steel and curse sank to the socket, as if sucked into a morass, Moby Dick sideways writhed; spasmodically rolled his nigh flank against the bow, and, without staving a hole in it, so suddenly canted the boat over, that had it not been for the elevated part of the gunwale to which he then clung, Ahab would once more have been tossed into the sea. As it was, three of the oarsmen—who foreknew not the precise instant of the dart, and were therefore unprepared for its effects—these were flung out; but so fell, that, in an instant two of them clutched the gunwale again, and rising to its level on a combing wave, hurled themselves bodily inboard again; the third man helplessly dropping astern, but still afloat and swimming.

Almost simultaneously the White Whale darted through the weltering sea. But when Ahab cried out to the steersman to take new turns with the line, and hold it so; and commanded the crew to turn round on their seats, and tow the boat up to the mark; the moment the treacherous line felt that double strain and tug, it snapped in the empty air!

"What breaks in me? Some sinew cracks!—'tis whole again; oars! oars! Burst in upon him!"

Hearing the tremendous rush of the sea-crashing boat, the whale wheeled round to present his blank forehead at bay; but in that evolution, catching sight of the nearing black hull of the ship; seemingly seeing in it the source of all his persecutions; bethinking it—it may be—a larger and nobler foe; of a sudden, he bore down upon its advancing prow, smiting his jaws amid fiery showers of foam.

Ahab staggered; his hand smote his forehead. "I grow blind; hands! stretch out before me that I may yet grope my way. Is't night?"

"The whale! The ship!" cried the cringing oarsmen.

"Oars! Oars! Dash on, my men! Will ye not save my ship?"

But as the oarsmen violently forced their boat through the sledge-hammering seas, the before whale-smitten bow-ends of two planks burst through, and in an instant almost, the temporarily disabled boat lay nearly level with the waves; its half-wading, splashing crew, trying hard to stop the gap and bail out the pouring water.

Meantime, for that one beholding instant, Tashtego's mast-head hammer remained suspended in his hand; while Starbuck and Stubb, standing upon the bowsprit beneath, caught sight of the down-coming monster just as soon as he.

"The whale, the whale! Up helm, up helm! Oh, all ye sweet powers of air, now hug me close! Let not Starbuck die, if die he must, in a woman's fainting fit. Up helm, I say—ye fools, the jaw! the jaw! Is this the end of all my bursting prayers? all my life-long fidelities? Oh, Ahab, Ahab, lo, thy work. Steady! helmsman, steady. Nay, nay! Up helm again! He turns to meet us! My God, stand by me now!"

"Stand not by me, but stand under me, whoever you are that will now help Stubb; for Stubb, too, sticks here. I grin at thee, thou grinning whale! Look ye, sun, moon, and stars! I call ye assassins of as good a fellow as ever spouted up his ghost. For all that, I would yet ring glasses with ye, would ye but hand the cup! Oh, oh! oh, oh! thou grinning whale, but there'll be plenty of gulping soon! Why fly ye not, O Ahab! For me, off shoes and jacket to it; let Stubb die in his drawers! A most mouldy and over-salted death, though;—cherries! cherries! cherries! Oh, Flask, for one red cherry ere we die!"

"Cherries? I only wish that we were where they grow. Oh, Stubb, I hope my poor mother's drawn my part-pay ere this; if not, few coppers will now come to her, for the voyage is up."

From the ship's bows, nearly all the seamen now hung inactive; hammers, bits of plank, lances, and harpoons, mechanically retained in their hands, just as they had darted from their various employments; all their enchanted eyes intent upon the whale. Retribution, swift vengeance, eternal malice were in his whole aspect, and

"Whale attack," 1845. *Russell and Purrington's panorama includes this depiction of the murderous sperm whale that broadsided and sank the Nantucket whaleship* Essex. *Well known to the public at the time, the story of the* Essex *was the first such record of a whale deliberately ramming a ship, an event that greatly impressed and influenced Melville.*

New Bedford Whaling Museum

spite of all that mortal man could do, the solid white buttress of his forehead smote the ship's starboard bow, till men and timbers reeled. Some fell flat upon their faces. Like dislodged trucks, the heads of the harpooneers aloft shook on their bull-like necks. Through the breach, they heard the waters pour, as mountain torrents down a flume.

"The ship! The hearse!—the second hearse!" cried Ahab from the boat; "its wood could only be American!"

Diving beneath the settling ship, the whale ran quivering along its keel; but turning under water, swiftly shot to the surface again, far off the other bow, but within a few yards of Ahab's boat, where, for a time, he lay quiescent.

"I turn my body from the sun. What ho, Tashtego! Let me hear thy hammer. Oh!—death-glorious ship! Must ye then perish, and without me? Am I cut off from the last fond pride of meanest shipwrecked captains? Oh, lonely death on lonely life! Oh, now I feel my topmost greatness lies in my topmost grief. Towards thee I roll, thou all-destroying but unconquering whale; to the last I grapple with thee; from hell's heart I stab at thee; for hate's sake I spit my last breath at thee. Sink all coffins and all hearses to one common pool! and since neither can be mine, let me then tow to pieces, while still chasing thee, though tied to thee, thou damned whale! *Thus*, I give up the spear!"

The harpoon was darted; the stricken whale flew forward; with igniting velocity the line ran through the groove;—ran foul. Ahab stooped to clear it; he did clear it; but the flying turn caught him round the neck and he was shot out of the boat, ere the crew knew he was gone. Next instant, the heavy eye-splice in the rope's final end flew out of the stark-empty tub, knocked down an oarsman, and smiting the sea, disappeared in its depths.

For an instant, the tranced boat's crew stood still; then turned. "The ship? Great God, where is the ship?" Soon they through dim, bewildering mediums saw her sidelong fading phantom, only the uppermost masts out of water; while fixed by infatuation, or fidelity, or fate, to their once lofty perches, the pagan harpooneers still maintained their sinking lookouts on the sea. And now, concentric circles seized the lone boat itself, and all its crew, and each floating oar, and every lance-pole, and spinning, animate and inanimate, all round and round in one vortex, carried the smallest chip of the Pequod out of sight.

But as the last whelmings intermixingly poured themselves over the sunken head of the Indian at the mainmast, leaving a few inches of the erect spar yet visible, together with long streaming yards of the flag, which calmly undulated, with ironical coincidings, over the destroying billows they almost touched;—at that instant, a red arm and a hammer hovered backwardly uplifted in the open air, in the act of nailing the flag faster and yet faster to the subsiding spar. A sky-hawk that tauntingly had followed the main-truck downwards from its natural home among the stars, pecking at the flag, and incommoding Tashtego there; this bird now chanced to intercept its broad fluttering wing between the hammer and the wood; and simultaneously feeling that ethereal thrill, the submerged savage beneath, in his death-gasp, kept his hammer frozen there; and so the bird of heaven, with archangelic shrieks, and his imperial beak thrust upwards, and his whole captive form folded in the flag of Ahab, went down with his ship, which, like Satan, would not sink to hell till she had dragged a living part of heaven along with her, and helmeted herself with it.

Now small fowls flew screaming over the yet yawning gulf; a sullen white surf beat against its steep sides; then all collapsed, and the great shroud of the sea rolled on as it rolled five thousand years ago.

Ahab meets Moby Dick. *"…To the last I grapple with thee; from hell's heart I stab at thee; for hate's sake I spit my last breath at thee.… Let me then tow to pieces, while still chasing thee, though tied to thee, thou damned whale! Thus, I give up the spear!"* — Chapter 135. Oil painting by Isaac Walton. Taber for Moby-Dick frontispiece; published by Charles Scribner's Sons, 1899.

Kendall Collection, NBWM

Bark Wanderer, circa 1920. Built in Mattapoisett in 1878, Wanderer was the last square-rigged whaleship to leave New Bedford. She made 23 voyages from New Bedford and San Francisco before being wrecked off Cuttyhunk Island in 1924.

Epilogue

"AND I ONLY AM ESCAPED ALONE TO TELL THEE." Job.

The Drama's Done. Why then here does any one step forth?—Because one did survive the wreck.

It so chanced, that after the Parsee's disappearance, I was he whom the Fates ordained to take the place of Ahab's bowsman, when that bowsman assumed the vacant post; the same, who, when on the last day the three men were tossed from out the rocking boat, was dropped astern. So, floating on the margin of the ensuing scene, and in full sight of it, when the half-spent suction of the sunk ship reached me, I was then, but slowly, drawn towards the closing vortex. When I reached it, it had subsided to a creamy pool. Round and round, then, and ever contracting towards the button-like black bubble at the axis of that slowly wheeling circle, like another ixion I did revolve, till gaining that vital centre, the black bubble upward burst; and now, liberated by reason of its cunning spring, and owing to its great buoyancy, rising with great force, the coffin-like buoy shot lengthwise from the sea, fell over, and floated by my side. Buoyed up by that coffin, for almost one whole day and night, I floated on a soft and dirge-like main. The unharming sharks, they glided by as if with padlocks on their mouths; the savage sea-hawks sailed with sheathed beaks. On the second day, a sail drew near, nearer, and picked me up at last. It was the devious-cruising Rachel, that in her retracing search after her missing children, only found another orphan.

Finis.

New Bedford Whaling Museum

General Index to Captions and Images

A

Adams, John 166
advertisements. *See whaling crew: advertisements for*
African Americans, in New Bedford 18. *See [also] Cape Verdeans*
Alaska (Kodiak) 76
Antarctica 36
Arctic whaling 40, 66, 69, 103, 105, 109, 130
Arnold, James 31
Arnold, Sarah Rotch 23
Arnold Arboretum 31
Azores Islands
    Fayal 69, 98
    Flores 69
    Pico 69, 98
    São Miguel 137

B

Bent, Joe, boatsteerer 69
Buru Island, Indonesia 140
Buzzards Bay, Massachusetts 44

C

calabooza beretanee (jail) 46
Cape Hatteras, North Carolina 69
Cape Horn 52, 164, 167, 176, 193
Cape of Good Hope 100
Cape Verdeans 59, 82, 99
Cape Verde Islands 98, 178, 196
    Brava 99
    Fago 99
    São João 99
Chase, Jack, seaman 47
Chase, Owen, First Mate 91
Chile, Isle of Mas Afuera, 84
Clark, Capt. Benjamin S. 15
Coleman, Capt. John 150
compass 191
Comstock, Samuel, mutineer 189
coopering 39
Crozet Islands 179
cutting-in. *See whale processing: cutting-in*
Cuttyhunk Island, Massachusetts
    Sow and Pig Reef 68

D

Dartmouth, Massachusetts 111
    Mishaum Point 44
Degrasse, Quinton, cabin boy 69
Dominican Republic 82
Dominique Island 192
donkey engine 62
"Down to the Sea in Ships," motion picture 22, 80, 94, 117, 143
Duxbury, Massachusetts 167

E

Edwards, Capt. Antone F. 68
Edwards, Capt. Joseph F. 69

F

Fairhaven, Massachusetts 13, 16, 19, 22, 40, 52, 101
    Fort Phoenix 14
    Rising Sun Inn 17
    Union Wharf 19, 40
Fernando Noronha Island 172
Fiji Islands 175
Freitas, Theophilo, harpooner 117

G

Galapagos Islands 52, 164
gamming 102-03, 192
Gay Head, Martha's Vineyard 223
Gibraltar 196
Gifford, Capt. Benjamin A. 137
Gonsalves, Capt. John 82
Greene, Tobias, seaman 46-47, 58, 192
Greenland whale. *See whale, types of: right or Greenland*

H

harpoons. *See whaling tools and implements: harpoons*
Haskins, Capt. Amos 69
Hawaiian Islands 148, 175
    Kealakekua Bay 175
    Lahaina Harbor, Maui 47, 175
Howland, George, Jr., merchant 55
Howland, W. F., Captain 40

J

Japan 67, 150
Japan Grounds, whaling on 149-50

L

Lewis, Frank, shipkeeper 61
lighthouses
    Clark's Point, New Bedford 44
    Sankaty Head, Nantucket 49
London, England 67, 84, 86, 131, 134, 141, 164-65, 167, 176, 187

M

Mandley, Capt. Henry S. 182
Manjiro, John 67
Maori people 25-26, 36-37
    carvings 26, 36
    chiefs 25, 36-37
    war canoe 37
Marquesas Islands 25, 46
    Hivaoa Island 149
    Typee Bay, Nuku Hiva 46-47
    Taiohae Bay 46
Massachusetts, steamer 44
Mattapoisett, Massachusetts 214
    Barstow Shipyard 52
Maury Whale Chart 90
Melville, Thomas 17
Morgan, Charles W., merchant 31
Mudge, Rev. Enoch, pastor 34
mutinies
    Globe, whaleship 189
    Sharon, whaleship 190
Mystic Seaport, Connecticut 53

N

Nantucket, Massachusetts 31, 48, 149-50, 166, 189
    Macy House 50
    Sankaty Head Light 49
    Sherbourne 48
Neves, Antão, harpooner 118
New Bedford, Massachusetts 13-14, 66, 109, 166-67
    Acushnet Heights 16
    alms house 44
    Bethel A.M.E. Church 18
    Central Wharf 51
    Clark's Point lighthouse 113
    County Street 31
    First Street 19
    Hazard's Wharf 220
    Johnny Cake Hill 23
    Liberty Hall 35
    Mariners' Home 23
    Merchant's Wharf 9, 38
    New Bedford Harbor 13-14, 40-45, 64, 109, 173
    Parker, John Avery, mansion 30
    Port Society 23, 33
    railroad depot 17
    Seamen's Bethel 32-34
    Second Street 19
    Taber's Wharf 38
    Union Street 19, 20, 28, 30
    Water Street 20, 28-29
New Zealand 24-25, 36-37, 125, 147
    Cape Brett 37
    Dunedin 125
    Hood's Island 192
    Maori people 25-26, 36-37
    Whangara 36-37
Norris, Capt. Howes 190
Nye, Capt. George 195
Nye, Thomas, merchant 55
Nye, William F., oil refinery 151

P

panoramas
    "Whaling Voyage in Ship Niger" 143
    "Whaling Voyage Round the World" 42-45, 47, 100, 137, 144, 149, 175-76, 192-93, 213
Pease, Jr., Capt. Valentine 46, 52
Peru 164
Phelone, Capt. Henry 49
Polynesian Islands 25, 190
Potter, Capt. Sylvester B. 197
Provincetown, Massachusetts 82, 131, 133
publications
    Harper's Magazine 66, 78
    Murphy's Journal 91
    Whalemen's Shipping List 58
    Wide Awake Library 15

Sperm whaling scene, 1840s. Scrimshaw on panbone (jawbone) plaque engraved by an anonymous whaleman-artist. — Kendall Collection, NBWM.

Q
quadrant 186
Quakers 22, 55

R
R. B. Forbes, tugboat 44
right whaling 111, 126, 129
Rio de Janeiro, Brazil 52
Rotch, Elizabeth 31
Rotch, Joseph 31
Rotch, William 31
Rotch, William, Jr. 23, 31
Russell, Capt. Charles C. 101

S
Sag Harbor, New York 84, 139
Samuel Enderby & Sons, merchants (London) 164-67, 171, 179
San Francisco, California 208
sea chest 173
shipyards 39, 52
    Barstow Shipyard 52
    Hillman Shipyard 53
ship and boat plans
    Alice Knowles, bark 172
    Beetle whaleboat 113
    Charles W. Morgan, whaleship 53
ship carpenters and caulkers 39, 52
Simoda, merchant vessel 69
Society Islands (Huahine) 175
South Sea whale fishery 21, 108, 140, 166-67
sperm whaling 106-08, 136-37, 145-48
storms 68, 187-88
stove boats 95, 109, 131, 201-02, 206
Straights of Sunda 167

T
Taber, Charles, publisher 107-08, 151
Taber, Nicholas, Innkeeper 17
Tahiti 25, 190
tattoo 25
taverns
    Rising Sun Inn 17
    sailors just landed 20
    Spouter Inn 22
    Swordfish Inn 40
Tisbury, Martha's Vineyard 15

U
U.S. Exploration Expedition 36
United States, frigate 47

W
Wampanoag 69
War of 1812 15
Webber, Frank, chandler 57
Westport, Massachusetts 146
whale, types of
    baleen 169
    blackfish 77
    blue or northern rorqual 168-69
    humpback 76
    lesser rorqual 169
    minke whale 169
    narwhale 77
    right or Greenland 76, 97, 104, 111, 126, 129-30
    skeletons 168-69
    sperm 75, 86, 94, 105, 114, 121, 124-25, 128, 131-33, 136-38, 140, 143, 145, 147, 167-70
    unicorn 77
whaleboats 86, 92, 109, 112-13, 124, 140, 150, 182, 207
whale anatomy and byproducts
    ambergris 125, 153
    baleen 129-30
    flukes 128, 143, 205
    head 124, 128
    jaw 88
    oil 132, 152-53
    spermacetti 124, 166
    sperm whale's yield 131
    teeth 110
Whale Point, Nova Scotia 194
whale processing
    bailing out the case 132, 156
    blanket pieces, cutting 122, 155
    carcass 123
    cutting-in 121-22, 125, 129, 133-34, 162-63
    drying baleen 122
    hauling the case 124, 132
    hoisting baleen 129
    stowing down oil 162
    stripping blubber 122, 155
    trying-out 151, 158-61
whaling conditions
    food 119
    life in the forecastle 78
    signing up 54
whaling crew
    aboard *Acushnet* (shipping paper) 58
    aboard *Baltic* 78
    aboard *Charles W. Morgan* 82
    aboard *John R. Manta* 119
    aboard *Orray Taft* 81
    advertisements for 15, 56
    bound out 64
    cook and the pilot 74
    individual seamen 69-70, 83, 116
    just landed 16
    masthead lookouts 80, 199
whaling ephemera / artifacts
    pulp fiction 15
    settlement receipts 57
    sperm whale oil bottles 132
    stationary 152
whaling scenes
    boats chasing and striking whales 93-96, 105, 115-118, 126, 130, 136-141, 144-51, 165, 184-85, 203, 205, 211
    dead whales 118, 130, 185
    lowering boats 92, 209
    pitchpoling 141
    taking up boats 135
    whales attacking boats 87, 142, 200, 213
    whales in a flurry 141-42, 205
whaling tools and implements 156
    blubber hooks 29, 122, 138
    coiling line into tubs 113
    harpoons
        Pierce's darting gun 181
        English harpoon 181
        explosive harpoon 181
        flued harpoon 181
        harpoon line 111
        lance harpoon 181
        strapping on 60
        Temple toggle iron 181
    head needle 124
    head spade 124
    sharpening a lance 180
    try pots 29
Whitfield, Capt. William H. 67, 172
Wilmington, Delaware 130, 144, 156

Index to Artists and Artisans

painters, illustrators

Agate, Alfred T. *26*
Ashley, Clifford W. *90, 123, 125, 154-55, 182*
Barber, John W. *7, 13*
Beest, Albert van *107-09*
Bradford, William *109, 208*
Brett, Oswald *49*
Ellis, Richard *88*
Gale, George A. *117*
Garneray, Ambroise Louis *106-07*
Gifford, R. Swain *108*
Hartwell, Alonzo *206*
Huggins, William John *108, 140*
Hulsart, Cornelius B. *148*
Kent, Rockwell *27, 71, 85, 200, 210*
Moniz, Arthur *203*
Norton, William E. *161-162*
Overend, William Heysman *139*
Polack, J.S. *147*
Purrington, Caleb *41*
Raleigh, Charles S. *102-03, 143*
Russell, Benjamin *41-45, 47, 64, 99, 100, 107-08, 135, 137, 144-45, 149, 175-76, 192, 213, 223*
Ryne, J. Van *100*
Snow, E. C. *95, 141*
Stewart, J. *97, 104, 138*
Strother, David Hunter *62, 78*
Swift, Clement Nye *128, 223*
Swift, R. G. N. *99*
Taber, Isaac Walton *212*
Vanderlyn, John *42*
Wall, William A. *16, 31*
Ward, John *166*
Weir, Jr., Robert Wallace *97*
Winn, J. R. *136*

artisans

Beetle, whaleboat builder *113*
Halsey, James, compass maker *191*
Maori carvings *26, 36*
Mendonça, Manoel E. de, seaman, (sea chest) *173*
Pierce, Ebenezer, inventor *181*
Temple, Lewis, blacksmith *17, 181*
Warren, Russell, architect *30*
Weeks, Ansel, master carpenter *52*

logbook/journal whaleman-artists

Akin, John Francis, ship *Virginia* *163, 196*
anonymous
  Arab, bark *101*
  Edward Harding, whaleship *115*
  Orray Taft, bark *81, 112*
Baker, Warren W., brig *Leonidas* *146*
Bertoncini, John, shipboard artist *149, 185*
Chase, Amos, Jr., ship *Constantine* *84, 197*
Chase, Capt. Joseph T., whaleship *Massachusetts* *47*
Dexter, Rodolphus W., bark *Chili* *15, 183, 209*
Gould, George A., ship *Columbia* *149-50, 184, 192*
Hayward, Henry C., bark *Awashonks* *133*
Hershey, Joseph Bogart, schooner *Esquimaux* *131, 133, 207*
Johnsons, Herny, whaleship *Acushnet* *50-51, 184*
Martin, John F., whaleship *Lucy Ann* *98, 130, 144, 156, 161*
Mott, William, ship *William* (London) *164*
Peters, Joseph, ship *Congaree* *131*
Scales, John C., bark *Pearl* *101, 170, 219*
Taylor, William W., whaleship *South Carolina* *111*
Wetling, Thomas, ship *William* (London) *84, 131, 134, 164-65, 167, 176, 187*
Wordel, David, bark *Canton* *142*

engravers, lithographers

Cole, J., engraver *107*
Hawkins, W., engraver *153*
Lane, Fitz Hugh, lithographer *14*
Lizars, W., engraver *97, 104, 138*
Maury, Lt. Matthew, cartographer *90*
Mayer, A., lithographer *154*
Perkins, Granville, engraver *205*
Read, W., engraver *147*
Roberts, W., engraver *201*
Vos, Marten de, engraver *35*

scrimshanders

anonymous pieces
  bird cage *171*
  corset busks *147, 150*
  pagoda drawer *110*
  pie-crust crimpers *110*
  panbone plaque *218*
  polychrome plaque *141*
  sperm whale teeth *15, 110*
  yarn winder *110*
Roderick, W. L., scrimshander *141*

photographers

Aldrich, Herbert L. *20, 69, 72, 130*
Ashley, Clifford W. *62, 124, 199*
"Down to the Sea in Ships," the film *22, 80, 94, 117, 143*
Earle, Capt. James *125*
Gifford, Pardon *177*
Huston, John, film director *33*
Packard, Arthur F. *33, 61, 63*
Tirrell, Joseph G. *29, 40, 55, 64, 180, 197-98, 208*
Tripp, William H. *113, 194*

authors

Beale, Thomas *75, 105, 108, 138*
Browne, J. Ross *184*
Bullen, Frank T. *161-62*
Dumont, Peter L. *205*
Macy, Obed *50*
Melville, Herman *7-10, 16-19, 22, 31, 34-36, 40, 43, 46-47, 67, 70, 75, 86, 105-08, 148, 150, 176, 178, 190, 192, 205, 213*
Scoresby, Willam *105*

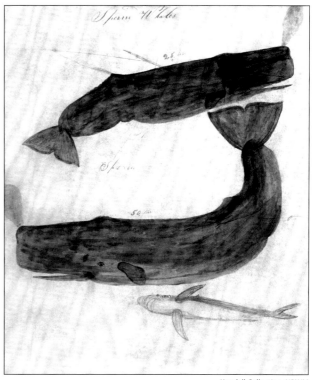

Sperm whales, 1852-54. —
Watercolor from the journal of John C. Scales aboard bark Pearl of New London.

Kendall Collection, NBWM

Index to Whaling Ships

A. R. Tucker, bark 197
Abraham Baker, bark 95, 141
Acushnet, whaleship 19, 40, 46, 52, 58, 91, 176, 178, 184, 192
Adam, whaleship (London) 167
Adventure, bark (London) 141
Alice Knowles, bark 57, 69, 172
Amelia Wilson, schooner 140
Andrew Hicks, bark 51
Ann Alexander, whaleship 86
Arab, bark 101
Atlantic, ship (England) 164
Awashonks, bark 133
Baltic, whaleship 78
Beale, whaleship 138
Blackfish, whaleship 77
Bonanza, schooner 185
Cachalot, whaleship 161-62
Canton, bark 38, 142, 177, 196
Castor, ship 140
Charles, whaleship 40
Charles and Henry, whaleship 47, 150
Charles W. Morgan, whaleship 53, 65, 73-74, 79-80, 82, 93, 98, 122, 125, 158, 160, 162
Chelsea, whaleship 121, 188
Chili, bark 15, 183, 209
Clara Bell, bark 97
Columbia, whaleship 149, 184, 192
Congaree, whaleship 131
Constantine, gold rush ship 84, 197
Contest, whaleship 99
Daisy, brig 92-93, 118
Desdemona, bark 194
Eagle, whaleship 192
Edward Harding, whaleship 115
Emilia, whaleship 167
Enterprise, brig 148
Esquimaux, schooner 131, 207
Essex, whaleship 86, 91, 213
Globe, whaleship 189
Gratitude, whaleship 45
Greyhound, bark 39, 198
Helen Mar, bark 160, 220
Houqua, whaleship 148
Huron, whaleship 84
India, bark 44
James Allen, bark 103
James Arnold, whaleship 208
Janus, whaleship 44
John J. Howland, whaleship 67
John R. Manta, schooner 60, 92, 111, 114, 119, 121, 122, 124, 132, 155, 177, 182, 185-86
Kutusoff, whaleship 43, 176
Lancer, bark 66
Leonidas, brig 146
Le Mercure, brig (France) 154
Lucretia, whaleship 72
Lucy Ann, whaleship (Sidney) 46-47, 150, 190
Lucy Ann, whaleship (Wilmington) 98, 130, 144, 156, 161
Malta, bark 183
Massasoit, bark 69
Morning Star, bark 39, 51
Niger, whaleship 44, 143, 208
Orray Taft, bark 81, 112
Pearl, bark 101, 170
Plantina, bark 195
Pocahontas, whaleship 148
President, bark 137
Rebecca, whaleship 167
Reindeer, bark 69
Samuel and Thomas, bark 133
Sea Fox, ship 103
Sharon, whaleship 41, 190
Shylock, schooner 133
Somers, brig 154
South Carolina, whaleship 111
Splendid, whaleship (N. Zealand) 125
Sullivan, brig 54
Sunbeam, bark 124, 154-56, 199
Swallow, bark 64, 198
Tamerlane, bark 40
Three Brothers, whaleship 49
Virginia, whaleship 163, 196
Wanderer, bark 61, 63, 68, 113, 116
William, ship (England) 84, 164, 167, 176, 187
William and Eliza, whaleship 172
William Mott, whaleship 134
William Rotch, whaleship 148
Young Phoenix, whaleship 69, 109, 201

Old Wharf Scene, 1866. Bark Helen Mar at Hazards Wharf (foreground), and other whalers lay over at New Bedford.

William R. Hegarty Collection

Bibliography

"A Summer in New England." Illustrations by Porte Cravon. *Harper's New Monthly Magazine*, June 1860.

Allen, Everett S. *Children of the Light; the Rise and Fall of New Bedford Whaling and the Death of the Arctic Fleet*. Boston: Little Brown, 1973.

Ames, Percy D. *The Building of the Whaleship Charles W. Morgan*. New Bedford, Mass.: Reynolds Printing, 1933.

Anthony, Joseph R., and Zephaniah W. Pease. *Life in New Bedford a Hundred Years Ago; a Chronicle of the Social, Religious and Commercial History of the Period as Recorded in a Diary Kept by Joseph R. Anthony*. New Bedford, Mass.: Old Dartmouth Historical Society, 1922.

Ashley, Clifford W., and Macbeth Gallery. *The Whalers of New Bedford: Exhibition of Paintings by Clifford W. Ashley, October 31st-November 13th 1916*. New York City: The Gallery, 1916.

Beale, Thomas. *The Natural History of the Sperm Whale... To Which Is Added, a Sketch of a South-Sea Whaling Voyage. In Which the Author Was Personally Engaged*. London: J. Van Voorst, 1839.

Bennett, Frederick Debell. *Narrative of a Whaling Voyage Round the Globe, from the Year 1833 to 1836. Comprising...the Natural History of the Climates Visited*. London: R. Bentley, 1840.

Blasdale, Mary Jean. *Artists of New Bedford: A Biographical Dictionary*. New Bedford, Mass.: The Old Dartmouth Historical Society, 1990.

Boss, Judith A., and Joseph D. Thomas. *New Bedford: A Pictorial History*. Norfolk, Virginia: Donning Co., 1983.

Bourne, Russell. *The View from Front Street: Travels through New England's Historic Fishing Communities*. 1st ed. New York: W.W. Norton, 1989.

Browne, J. Ross. *Etchings of a Whaling Cruise, with Notes of a Sojourn on the Island of Zanzibar. To Which Is Appended a Brief History of the Whale Fishery*. New York: Harper, 1846.

Bullen, Frank Thomas. *The Cruise of the Cachalot, Round the World after Sperm Whales*. New York: D. Appleton Co., 1899.

Cheever, Henry T. *The Island World of the Pacific; Being... Travel through the Sandwich or Hawaiian Islands and Other Parts of Polynesia*. New York: Harper & Brothers, 1851.

Cheever, Henry T., and William Scoresby. *The Whaleman's Adventures in the Southern Ocean; as Gathered...on the Homeward Cruise of the "Commodore Preble."* London: Darton, 1861.

Church, Albert Cook. *Whale Ships and Whaling*. New York: W. W. Norton & Co., 1938.

Comstock, William. *The Life of Samuel Comstock, the Terrible Whaleman: Containing an Account of the Mutiny, and Massacre of the Officers of the Ship Globe...* Boston: James Fisher, 1840.

Creighton, Margaret S. *Rites and Passages: The Experience of American Whaling, 1830-1870*. Cambridge, England; New York, USA: Cambridge University Press, 1995.

Creighton, Margaret S., and Peabody Museum of Salem. *Dogwatch and Liberty Days: Seafaring Life in the Nineteenth Century*. Salem, Mass.: Peabody Museum of Salem, 1982.

Dahl, Curtis. "Of Foul Weather and Bulkingtons." *Melville Society Extracts*, May 1977, p. 10.

Dakin, William J. *Whalemen Adventurers; the Story of Whaling in Australian Waters and Other Southern Seas Related Thereto, from the Days of Sails to Modern Times*. Sydney, Australia: Angus & Robertson, 1938.

Davis, William M. *Nimrod of the Sea; or, the American Whaleman*. New York: Harper & Brothers, 1874.

Delano, Reuben. *Wanderings and Adventures of Reuben Delano: Being a Narrative of Twelve Years Life in a Whale Ship: Now First Published. Illustrated with Engravings*. Worcester, Mass.: Thomas Drew, Jr., 1846.

Dickerman, Marion. *The Story of the Last of the Old Whalers: "Charles W. Morgan"*. Mystic, Conn.: Marine Museum of the Marine Historical Association Inc., 1949.

Dow, George Francis. *Whale Ships and Whaling; a Pictorial... with an Account of the Whale Fishery in Colonial New England*. New York: Argosy Antiquarian, 1967.

Earle, James A. M. *Fighting Sperm Whales*. New York, 1927.

Ellis, Leonard Bolles. *History of New Bedford and Its Vicinity, 1620-1892*. Syracuse, N.Y.: D. Mason & Co., 1892.

Flayderman, Norm. *Scrimshaw and Scrimshanders: Whales and Whalemen*. New Milford, Conn.: N. Flayderman, 1972.

Frank, Stuart M., and Herman Melville. *Herman Melville's Picture Gallery: Sources and Types of the "Pictorial" Chapters of Moby-Dick*. Fairhaven, Mass.: E.J. Lefkowicz, 1986.

Georgianna, Daniel, and Roberta Hazen Aaronson. *The Strike of '28*. New Bedford, Mass.: Spinner Publications, 1993.

Gifford, Pardon B., and Zephaniah W. Pease. *100th Anniversary of the New Bedford Mercury, 1807 to 1907*. New Bedford, Mass.: Mercury Publishing Co., 1907.

Gifford, Pardon B., and William F. Williams. *The Wing Fleet. The Stone Fleet*. New Bedford, Mass.: Reynolds Printing, 1925.

Hall, Elton Wayland. *Sperm Whaling from New Bedford: Clifford W. Ashley's Photographs of Bark Sunbeam in 1904*. New Bedford, Mass.: Old Dartmouth Historical Society, 1982.

_____. *American Maritime Prints: The Proceedings of the Eighth Annual North American Print Conference...* New Bedford, Mass.: Old Dartmouth Historical Society, 1985.

Hamilton, Robert, and W. H. Lizars. *The Natural History of the Ordinary Cetacea or Whales*. Edinburgh: W.H. Lizar, et. al., 1837.

"Taking a whale," 1846. — *Frontispiece from J. Ross Browne.*

Hegarty, Reginald B. *Returns of Whaling Vessels Sailing from American Ports. A Continuation of Alexander Starbuck's "History of the American Whale Fishery," 1876-1928*. New Bedford, Mass.: Old Dartmouth Historical Society, 1959.

Hegarty, Reginald B., Walter E. Channing, and Milton Kenneth Delano. *Birth of a Whaleship*. New Bedford, Mass.: New Bedford Free Public Library, 1964.

"Herman Melville: The Symbolic Writer." *MD Magazine*, December 1969.

"Herman Melville: Moby-Dick." In *The Great Writers*. London: Marshall Cavendish, 1986-87.

Hirshson, G. Warren. *The Whale Ship Charles W. Morgan*. New Bedford, Mass.: Reynolds Printing Co., 1932.

Howland, Chester S. *Thar She Blows! Whaling Stories*. New York: Funk, 1951.

Huse, Donna, and Joseph D. Thomas. *Spinner: People and Culture in Southeastern Massachusetts*. Vol. II, *"Twentieth Century Whaling Tales."* New Bedford, Mass.: Spinner Publications, 1982.

Ingalls, Elizabeth. *Whaling Prints in the Francis B. Lathrop Collection*. Salem, Mass.: Peabody Museum of Salem, 1987.

Jones, John B. *Life and Adventure in the South Pacific by a Roving Reporter*. New York: Harper & Brothers, 1861.

Kendall Whaling Museum, M. V. Brewington, and Dorothy E. R. Brewington. *Kendall Whaling Museum Prints*. Sharon, Mass.: Kendall Whaling Museum, 1969.

Kugler, Richard C. "Herman Melville and New Bedford." *Sunday Standard-Times*, Nov. 30, 1969, p. 49.

_____. *William Allen Wall, an Artist of New Bedford: An Exhibition Held in Celebration of the Seventy-Fifth Anniversary of the Old Dartmouth Historical Society....* New Bedford, Mass.: Old Dartmouth Historical Society, 1978.

Leavitt, John F. *The Charles W. Morgan*. Mystic, Conn.: Mystic Seaport, 1973.

Leyda, Jay. *The Melville Log; a Documentary Life of Herman Melville, 1819-1891*. Two Volumes. New York: Gordian Press, 1969.

Martin, Kenneth R., and Kendall Whaling Museum. *Whalemen's Paintings and Drawings: Selections from the Kendall Whaling Museum Collection*. Sharon, Mass.: Kendall Whaling Museum 1983.

McCabe, Marsha, and Joseph D. Thomas. *Portuguese Spinner: An American Story*. New Bedford, Mass.: Spinner Publications, 1998.

Melville, Herman. *Typee: A Peep at Polynesian Life: During a Four Months' Residence in a Valley of the Marquesas*. New York: Wiley and Putnam, 1847.

_____. *Moby-Dick or The Whale*. New York: Harper and Brothers, 1852.

Melville, Herman, and Rockwell Kent. *Moby Dick or the Whale*. New York: Modern Library, 1930.

Murphy, Robert Cushman. *A Dead Whale or a Stove Boat; Cruise of Daisy in the Atlantic Ocean, June 1912-May 1913*. Boston: Houghton Mifflin Co., 1967.

Murray, Lt. Matthew F., U.S. Navy. "Whale Chart (Preliminary Sketch). Series F of Wind and Current Charts." Washington: Dept. of U.S. Navy, 1851.

New Bedford Board of Trade, Zephaniah W. Pease, George A. Hough, and William L. Sayer. *New Bedford, Massachusetts: Its History, Industries, Institutions and Attractions*. New Bedford, Mass.: Mercury Publishing Co., 1889.

New Bedford Free Public Library. *A Collection of Books, Pamphlets, Logbooks, Pictures, Etc. Illustrating the Whale Fishery, . . .* New Bedford, Mass., 1907.

Olmsted, Francis Allyn. *Incidents of a Whaling Voyage to Which Are Added Observations. . . of the Sandwich and Society Islands*. New York: D. Appleton, 1841.

Peabody Museum of Salem., M. V. Brewington, and Dorothy E. R. Brewington. *The Marine Paintings and Drawings in the Peabody Museum*. Rev. ed. Salem: Peabody Museum of Salem, 1981.

Pease, Zephaniah W. *The Story of the Celebration of the Semi-Centennial of the Incorporation of New Bedford as a City*. New Bedford, Mass.: Mercury Publishing Co., 1897.

_____. *History of New Bedford*. New York: Lewis Historical Publishing Co., 1918.

Pease, Zephaniah W., Frank Wood, and Old Dartmouth Historical Society. *The Arnold Mansion and Its Traditions*. New Bedford, Mass.: Vining Press, 1924.

Polack, Joel Samuel. *New Zealand: Being a Narrative of Travels and Adventures During a Residence in That Country between the Years 1831 and 1837*. Christchurch, New Zealand: Capper Press, 1974.

Purrington, Philip E. "Around the World in Eighty Rods: New Bedford's Whaling Panorama." *Antiques*. New York: 1965.

_____. "Anatomy of a Mutiny, Ship Sharon, 1842." *Old Dartmouth Historical Sketches, No. 75*. New Bedford, Mass.: Old Dartmouth Historical Society, 1968.

Robertson-Lorant, Laurie. *Melville: a Biography*. New York: Clarkson Potter, 1996.

Rodman, Samuel, and Zephaniah W. Pease. *The Diary of Samuel Rodman; a New Bedford Chronicle of Thirty-Seven Years, 1821-1859*. New Bedford, Mass.: Reynolds Printing Co., 1927.

Ronnberg, Erik A.R., Jr. "Photographs of Whaling Vessels: A Pictorial Supplement of the American Neptune." In *Old Dartmouth Historical Sketches*. New Bedford, Mass.: Old Dartmouth Historical Society, 1973.

Schultz, Elizabeth A. *Unpainted to the Last: Moby-Dick and Twentieth-Century American Art*. Lawrence, Kan.: University Press of Kansas, 1995.

Scoresby, William. *The Arctic Regions, and the Northern Whale-Fishery*. Vol. vii. London: The Religious Tract Society, 1849.

Sleeper, John Sherburne. *Tales of the Ocean, and Essays for the Forecastle: Containing Matters and Incidents Humorous, Pathetic, Romantic, and Sentimental*. Boston: G. W. Cottrell, 1857.

Starbuck, Alexander. *History of the American Whale Fishery, from Its Earliest Inception to the Year 1876*. New York: Argosy-Antiquarian, 1964.

Story, Dana. *The Shipbuilders of Essex: A Chronicle of Yankee Endeavor*. Gloucester, Mass.: Ten Pound Island Book Co., 1995.

Massachusetts Institute of Technology. "The Forbes Collection of Whaling Prints at the Francis Russell Hart Nautical Museum." Edited by Arthur C. Watson. Cambridge, Mass: M.I.T., 1941.

Thomas, Joseph D., and Marsha McCabe. *Spinner: People and Culture in Southeastern Massachusetts*. Vol. IV. New Bedford, Massachusetts: Spinner, 1988.

Townsend, Charles Haskins. "The Distribution of Certain Whales as Shown by Logbook Records of American Whaleships." *Zoologica,* 19, No. 1. New York: New York Zoological Society, 1935.

Tripp, William H. *"There Goes Flukes!"* New Bedford, Mass.: Reynolds Printing, 1938.

Vincent, Howard Paton. *The Trying-out of Moby-Dick*. Boston: Houghton Mifflin Co., 1949.

Weaver, Raymond M. *Herman Melville, Mariner and Mystic*. New York: George H. Doran Co., 1921.

"Fast to a whale," 1900. — *Sketch by Clement Nye Swift. Kendall Collection, NBWM.*

Whalemen's Shipping List, and Merchant's Transcript. March 19, 1850. New Bedford, Mass.: Benjamin Lindsey.

_____. Vol. 72., No. 52. Dec. 29, 1914. New Bedford, Mass.: Benjamin Lindsey.

Whitman, Nicholas. *A Window Back: Photography in a Whaling Port*. New Bedford, Mass.: Spinner Publications, 1994.

Wilkes, Charles. *Narrative of the United States Exploring Expedition During the Years 1838, 1839, 1840, 1841, 1842*. Philadelphia: Lea and Blanchard, 1845.

Wood, Edmund, and William A. Wing. *Address, Benjamin Russell*. New Bedford, Mass.: Old Dartmouth Historical Society, 1911.

Work Projects Administration, The Survey of Federal Archives Division of Professional and Service Projects. "Ship Registers of New Bedford, Massachusetts. Volume I, 1796-1850." Boston: The National Archives, 1940.

"Nearing Home," circa 1865. *A New Bedford whaling ship is close to home. In the background, at left, is Gay Head, Martha's Vineyard; to the right is Nomans Island.* — *Watercolor by Benjamin Russell.*

Forbes Collection of the MIT Museum

Spinner Publications, Inc. · New Bedford, Massachusetts

For information about Spinner books, calendars and historic photographs,
visit www.spinnerpub.com or call 800-292-6062